Artificial Intelligence in the Energy Industry

Artificial Intelligence in the Energy Industry

Editor

Ana-Belén Gil-González

MDPI • Basel • Beijing • Wuhan • Barcelona • Belgrade • Manchester • Tokyo • Cluj • Tianjin

Editor
Ana-Belén Gil-González
BISITE Research Group,
University of Salamanca,
Edificio Multiusos I+D+i,
Calle Espejo 2,
37007 Salamanca, Spain

Editorial Office
MDPI
St. Alban-Anlage 66
4052 Basel, Switzerland

This is a reprint of articles from the Special Issue published online in the open access journal *Energies* (ISSN 1996-1073) (available at: https://www.mdpi.com/journal/energies/special_issues/Artificial_Intelligence_Energy_Industry).

For citation purposes, cite each article independently as indicated on the article page online and as indicated below:

LastName, A.A.; LastName, B.B.; LastName, C.C. Article Title. *Journal Name* **Year**, *Volume Number*, Page Range.

ISBN 978-3-0365-4605-6 (Hbk)
ISBN 978-3-0365-4606-3 (PDF)

Cover image courtesy of Ana-Belén Gil-González.

Contents

About the Editor

Ana-Belén Gil-González

Ana-Belén Gil-González is Associate Professor at the Department of Computer Science and Automatics at the University of Salamanca in the area of Computer Languages and Systems. She holds a PhD in Computer Science and Automatics from the University of Salamanca, where she was awarded in recognition of her extraordinary PhD work. She currently holds the position of Vice-Dean of Teaching and Infrastructures of the Faculty of Science.

She works in the research group Bioinformatics, Intelligent Information Systems and Educational Technology (BISITE). Her research focuses on the development of technological frameworks for content retrieval, personalization and characterization in different domains and deepening its representation, analysis and application of content. In these works, she applied artificial intelligence techniques, multi-agent systems, data mining and Semantic Web for the retrieval and representation of knowledge, as well as its application to different systems such as Recommender Systems, as well as in applications related to Smartcities. A.B. Gil works on projects in the fields of Artificial Intelligence, Machine Learning, Blockchain, IoT, Fog Computing, Edge Computing, Smart Cities, Smart Grids, Sentiment Analysis, etc.

Her extensive experience with publication and dissemination of results is highlighted by 44 publications in international journals included in the Web of Science (WOS), of which 18 have an impact factor, according to the JCR-SCI index. She is also co-author of more than 100 publications in books and international peer-reviewed congresses of recognized prestige, some of them included in the CORE and SCI rankings, among others. She has collaborated in various national and international scientific committees while being active in the organization of numerous international scientific congresses (PAAMS, CEDI, DCAI, etc.).

Preface to "Artificial Intelligence in the Energy Industry"

Artificial intelligence is essential in all industrial environments. The energy industry is an area that presents exceptional opportunities for development based on the use of artificial intelligence (AI). In essence, AI provides a machine with the ability to learn and make decisions to solve problems or optimize results toward meeting a goal. There are many decisions to be made in the energy sector that require an early response and handling of a significant volume of data. Artificial intelligence can optimally perform these important decisions where instantaneous collection and analysis of these large volumes of data is required while processing as fast and efficiently as possible.

Smart grids carry electricity as well as data. In the case of intermittent and volatile energies, such as solar and wind, it is more important than ever to effectively balance consumption and generation. One of the hopes for artificial intelligence applied to the energy sector is that it will help us address issues related to climate change, emission-reduction effects of technological progresses in industry, energy balances, and environmental impacts, among others. One of the most basic applications of AI in the energy sector is the use of machine learning in making generation systems more efficient, improving the effectiveness of design technologies and creating energy-efficient objects.

The future of mobility is electric, which also poses new challenges. AI is being installed in the electric vehicle sector within cars themselves for their management and to communicate data that contribute to solving these challenges, in addition to outside the car to facilitate the effective management of reports, intelligent mobility solutions, etc.

The application of AI to the energy industry sector is without a doubt unquestionable. Artificial intelligence is beginning to be used in the energy sector and is already proving essential by providing the industry and households with new information services for control over energy infrastructure, optimizing generation, reducing consumption, or fighting climate change, which are only some examples of its promising applications that are expected in the near future.

This book showcases all the various research approaches focused on the relationship between the use of artificial intelligence and its direct application in the field of the energy sector, or so-called smart energy. The different chapters address the high incidence of contributions from the perspective of energy efficiency in energy management, production, and consumption. The use of AI will therefore make it possible to produce, consume, and manage energy and energy products better, with fewer resources and less environmental impact.

Ana-Belén Gil-González
Editor

 *energies*

Article

Comparison of Heat Demand Prediction Using Wavelet Analysis and Neural Network for a District Heating Network

Szabolcs Kováč *, German Micha'čonok, Igor Halenár and Pavel Važan

Institute of Applied Informatics, Automation and Mechatronics, Faculty of Materials and Science and Technology in Trnava, Slovak University of Technology in Bratislava, 917 02 Trnava, Slovakia; german.michalconok@stuba.sk (G.M.); igor.halenar@stuba.sk (I.H.); pavel.vazan@stuba.sk (P.V.)
* Correspondence: szabolcs.kovac@stuba.sk

Abstract: Short-Term Load Prediction (STLP) is an important part of energy planning. STLP is based on the analysis of historical data such as outdoor temperature, heat load, heat consumer configuration, and the seasons. This research aims to forecast heat consumption during the winter heating season. By preprocessing and analyzing the data, we can determine the patterns in the data. The results of the data analysis make it possible to form learning algorithms for an artificial neural network (ANN). The biggest disadvantage of an ANN is the lack of precise guidelines for architectural design. Another disadvantage is the presence of false information in the analyzed training data. False information is the result of errors in measuring, collecting, and transferring data. Usually, trial error techniques are used to determine the number of hidden nodes. To compare prediction accuracy, several models have been proposed, including a conventional ANN and a wavelet ANN. In this research, the influence of different learning algorithms was also examined. The main differences were the training time and number of epochs. To improve the quality of the raw data and remove false information, the research uses the technology of normalizing raw data. The basis of normalization was the technology of the Z-score of the data and determination of the energy-entropy ratio. The purpose of this research was to compare the accuracy of various data processing and neural network training algorithms suitable for use in data-driven (black box) modeling. For this research, we used a software application created in the MATLAB environment. The app uses wavelet transforms to compare different heat demand prediction methods. The use of several wavelet transforms for various wavelet functions in the research allowed us to determine the best algorithm and method for predicting heat production. The results of the research show the need to normalize the raw data using wavelet transforms. The sequence of steps involves following milestones: normalization of initial data, wavelet analysis employing quantitative criteria (energy, entropy, and energy-entropy ratio), optimization of ANN training with information energy–entropy ratio, ANN training with different training algorithms, and evaluation of obtained outputs using statistical methods. The developed application can serve as a control tool for dispatchers during planning.

Keywords: artificial neural networks; data analysis; signal decomposition; district heating; forecasting

Citation: Kováč, S.; Micha'čonok, G.; Halenár, I.; Važan, P. Comparison of Heat Demand Prediction Using Wavelet Analysis and Neural Network for a District Heating Network. *Energies* **2021**, *14*, 1545. https://doi.org/10.3390/en14061545

Academic Editor: Antonio Rosato

Received: 18 January 2021
Accepted: 8 March 2021
Published: 11 March 2021

Publisher's Note: MDPI stays neutral with regard to jurisdictional claims in published maps and institutional affiliations.

1. Introduction

The application of new, progressive technologies to process control is the key to increasing productivity, quality, reliability, and safety [1]. The use of modern process control means predictive control in the management process. It can be achieved by the implementation of artificial intelligence in the control process, together with other data processing methods. This article deals with the problem of short-time (1 h ahead) prediction of energy consumption and planning in heat production, with a comparison of the effectiveness of different methods and algorithms.

Planning is the first phase and is often the most important in many areas. In the area of heat production, the amount of energy produced depends on several variables.

The most economical solution would be to produce as much as the demand. This is difficult to achieve in heating systems due to many factors. First, to provide enough heat to the delivery point, it is necessary to produce more heat than is needed. For maximal production efficiency, it is important to make correct decisions. It is convenient to be able to forecast future states and make predictive decisions. The ability to predict is very beneficial generally, such as in systems with high inertia. The control system of a thermal power plant has large inertia in response to changes in the load on the part of customers and will therefore require accurate prediction of the production of the required power and its distribution to individual positions. The current trend in the heating industry is the transition from the third generation to the fourth generation which is characterized by high efficiency, low heat losses, renewable and excess energy utilization. The fourth generation district heating system (DHS) also presents concept that uses low supply temperature which significantly reduces DHS inertia. Transition to the fourth generation can be achieved through the intelligent design of the network—implementing smart meters, smart forecasting algorithm, etc. [2]. District heating is mainly influenced by the demand that means accurate demand prediction is necessary. However, in order to ensure the accuracy of the prediction, it is necessary to include the energy required by the consumers. At present, the accuracy of the predicted behavior of energy consumers is provided by mathematical models, operational statistics, and the experience of dispatchers. This places great demand on dispatcher skills in the production planning process. The use of artificial intelligence opens up new possibilities for improving the accuracy of thermal power plant management and thus increasing the efficiency of energy production. Finally, the improvement of heat load prediction accuracy can ensure the comfort of users and improve energy utilization. The central heating sector can play a significant role in reducing emissions [3] and has an influence on global climate change. It is, of course, also important to consider the availability, environmental friendliness, and price of fuels and the heat transfer medium [4]. The heat consumption originates mainly from heat used for space heating and tap water heating. This combination leads to nonlinearity. It is confirmed that neural networks have ability to solve nonlinearity, and their implementation is successfully proven in many cases and described in many publications [5–7].

Besides neural networks, it is possible to use many other methods for short-term prediction. One of them is an application of wavelet transform to analyze data from the predicted process. The aim of this article is a comparison of the prediction accuracy of the mentioned methods. The source was data from the real environment of the heat production company for 2014-2015, with 2018 used for validation. The local heat plant uses the software TERMIS to optimize heat demand, and the control of the heating process is based on mathematical modeling and statistical methods. Although the data from the environment and heat production are stored in the database, the control of the heating process does not use any method of prediction and is based on mathematical calculations.

The first step is to choose a proper neural network for the creation of the prediction model using data from the heating process. The next steps are the creation of a selected neural network, identifying the data that will be intended for training, testing, and valida-tion, and realization of the learning process itself. After these steps, it is possible to create a software module for predictive control of the heating process.

Similarly, in the second part of this research, the wavelet transform is applied to raw data. First, it is very important to choose a proper method of wavelet transform. Therefore, several wavelet transformations will be computed for different wavelet functions. After these steps, it is possible to make a comparison of different prediction methods for the control of the heating process to choose the best algorithm, whether a neural network or a combination of wavelet transform and neural network.

2. Related Work and Theoretical Basis

This section presents a brief description of the Discrete Wavelet Transform (DWT) and Artificial Neural Networks (ANN) that are used in this research.

2.1. Wavelet Theory and Multiresolution Analysis

Thanks to progress in computer science, there are many cases in which data from time-dependent processes in the physical world were processed using a computer system [8]. There are many algorithms for the better understanding of these processes. The best known and most used is Fourier transform (FT). For computer processing, FT is often paired with Gabor transform, S-transformation, Hilbert transform, or Wavelet transform. In addition to the previously mentioned transformation methods, empirical mode decomposition (EMD) or ensemble empirical mode decomposition (EEMD) are used to solve similar signal processing and surface reconstruction problems [9–11]. Wavelet transform and EMD/EEMD are relatively new. The main scope of this contribution is to use wavelet transformation for signal processing, prediction, and to improve the resolution of the data from the controlled process.

Thanks to the work of Meyer and Mallatt [12,13], wavelets have become widely known. Wavelet transform, thanks to its properties, is usable in many fields—mainly picture and video processing [14–17], fault detection [18–20], diagnostics and research in medicine [21,22], but also in many other fields [23–25].

Within the management of production processes, a common mathematical task is the prediction of the future state of a system based on known, recorded data. The wavelet transform is not mainly intended as a forecasting technique. It transforms a suspected signal into different levels of resolution and localizes a process in time and frequency. Even so, there are a lot of papers in which it is used in the prediction process.

As we can see from the available literature, wavelet transformation is often used with other technologies, mainly in the process of prediction based on a known system state, data, and parameters from the past. The methods used for the computation of future state differ, including different types of neural network, Nonlinear Least Squares Autoregressive Moving Average (NLS-ARMA) [23,25], high-dimension space mapping [26,27], fractal prediction [28], grey prediction [29], etc. For example, Elarabi et al. [30] proposed an approach that uses both DCT (Discrete Cosine Transform) and DWT (Discrete Wavelet Transform) to enhance the intraprediction phase of H.264/AVC standard. The algorithm presented in their research is designed for video processing software and, according to the authors, extends the benefits of the wavelet-based compression technique to the speed of the FSF algorithm and forms an intraprediction algorithm that ensures a 51% drop in bit rate while keeping the same visual quality and peak signal-to-noise ratio (PSNR) of the original H.264/AVC intraprediction algorithm. A very interesting approach joining wavelet decomposition and adaptive neuro-fuzzy inference system (ANFIS) for ship roll forecasting is described in the work of Li et al. [31].

Stefenon et al. [32] presented an approach to predict the failure of insulators in electric power systems. In their work, they used a hybrid approach with wavelets and neural networks to process data from the ultrasonic scan. Another use of the wavelet method is described in the work of Prabhakar et al. [33]. They describe a combination of Fast Fourier transform (FFT) together with Discrete Wavelet Transform and Discrete Shearlet Transform for predicting surface roughness by milling. The predicted results of the hybrid model were better than the individual transform. The combination of wavelet transform and neural network for prediction is described by Zhang et al. [34]. The core of the article is the creation of a power forecasting model based on dendritic neuron networks in combination with wavelet transform. For decomposing input data in the proposed model, Mallat's algorithm, which is a fast Discrete Wavelet Transform (DWT), is used. The results show that, with the help of wavelet decomposition, together with various types of neural networks, the prediction process is faster and better compared to the results obtained by the other three conventional models for almost every error criterion.

Another case of joining neural networks with DWT is described in [35]. The article describes the proposal of a hybrid system for prediction that consists of a Long Short-Term Memory (LSTM) neural network and a wavelet module. Wavelet transform is used to decompose the data into a set of subseries, which appears to be very effective. El-Hendawi and Wang [36] describe using a full wavelet packet transform model together with a neural network. Their proposed model is able to predict the electrical load. The model consists of a wavelet packet transform module that is able to decompose a series of high-frequency components into subseries of high and low frequencies. Subseries are fed into the neural networks and the outputs of each neural networks are reconstructed, which is the forecasted load. The described approach is not sensitive to various conditions such as different day types (e.g., weekend, weekday, or holiday) or months. The authors also state that the model can reduce MAPE error by 20% compared to the conventional approach.

Similar problems are solved in [37–39]. Tayab et al. [39] propose using wavelet transform with classical feed-forward neural network for short-term forecasting of electricity load demand. Prediction in their work consists of using the best-basis stationary wavelet packet transform. The authors used a Harris hawks optimization to optimize the feed-forward neural network weights. The hybrid model achieved a more than 60% decrease in MAPE compared to SVM and a classical backpropagation neural network. More sophisticated methods are described by Liu et al. [37]. The article predicts wind speed in a wind power generation plant. The highlight is a conjunction of advanced neural network models with wavelet packet decomposition (WPD). The authors developed a new hybrid model to predict wind speed. The model is based on a WPD, convolutional neural network (CNN), and convolutional long short-term memory network (CNNLSTM). In the developed WPD-CNNLSTM-CNN model, the WPD is used to decompose the original wind speed time series into various subseries. CNN with a 1D convolution operator is employed to predict the obtained high-frequency subseries and CNNLSTM is employed to complete the prediction of the low-frequency subseries.

The same or a very similar approach to signal processing via wavelet transformation is used in the work of Farhadi et al. [40]. They used a proven procedure of data processing, like other authors. This means that data from the manufacturing process retrieved via piezoelectric sensors and microphones are processed with a combination of wavelet transform and neural network. All calculations are processed by a program written in MATLAB software. The type of neural network used is multilayer perceptron with a backpropagation learning method.

Moreover, wavelet transform, in combination with other algorithms, can be used to filter out noise or interference signals on the premise of ensuring an undistorted original and adding the data prediction function in the data denoising process. Feng et al. [41] described a wavelet-based Kalman smoothing approach for oil well testing during the data processing stage. To improve the dynamic prediction and data resolution, similar approaches with the combination of Kalman prediction and wavelet transform can be found in [42,43]. There is the possibility to use a combination of DWT and other data filters. For example, the most often used method is the common and simple moving average (MA) method [44,45].

It is clear that the wavelet method (DWT) can be widely used for computer data processing in many applications and many scientific fields, together with different technologies, but mainly neural networks. Theory from the area of wavelet transformation is well described in many works [46–49]. The DWT is considered a linear transformation for which wavelets are discretely sampled. Multiresolution analysis (MRA) and filter bank reconstruction are properties that confirm the wide range of applicability of DWT [50]. The basis functions are derived from a mother wavelet $\psi(x)$, by factors of dilation and translation [51].

$$\psi_{a,b}(x) = \frac{1}{\sqrt{a}}\psi\left(\frac{x-b}{a}\right) \tag{1}$$

where a is the dilation factor and b represents the translation factor. The continuous wavelet transform of a function $f(x)$ can be expressed as follows:

$$F_w(a,b) = \frac{1}{\sqrt{a}}\int_{-\infty}^{\infty} f(x)\psi^*\left(\frac{x-b}{a}\right)dx \tag{2}$$

where * represents the conjugate operator. The basis functions in (Equation (1)) are redundant when a and b are continuous, so it is possible to discretize a and b to form an orthonormal basis. One way of discretizing a and b is to let $a = 2^p$ and $b = 2pq$. After this, (Equation (2)) can be expressed as follows:

$$\psi_{a,b}(x) = 2^{-p/2}\psi^*\left(2^{-p}x - q\right) \tag{3}$$

where p and q are integers. Then, the wavelet transform in (Equation (3)) can be expressed as follows:

$$F_w(a,b) = 2^{-p/2}\int_{-\infty}^{\infty} f(x)\psi^*\left(2^{-p}x - q\right)dx \tag{4}$$

where p and q are set to be integers, and it is possible to call (Equation (2)) a wavelet series. It is clear from the representation that the wavelet transform contains both spatial and frequency information. The wavelet transform is based on the concept of multiresolution analysis. That means the signal is decomposed into a series of subsignals and their associated detailed signals at different resolution levels. Generally, these subseries are called approximations (low frequencies) and details (high frequencies). The smooth subsignal (approximation) at level m can be reconstructed from the ith level smooth subsignal and the associated $m + 1$ detailed signals.

Matlab software is often used for mathematical analysis [52]. Thanks to the number of functions and features, it is suitable for wavelet decomposition, too. The wavelet problem is well managed by the wavelet toolbox.

The principle of the DWT algorithm is depicted in Figure 1, where, after employing the DWT, the procedure comprises of log2 N steps. From signal s, the very first step is to produce two sets of coefficients: approximation coefficients cA1 and detail coefficients cD1. After convolution of the signal s with the lowpass filter LoD and the highpass filter HiD, the dyadic transformation (downsampling) is applied. After obtaining approximation cA1 and detail cD1, the procedure of downsampling continues until the condition of N steps is met. The downsampling procedure is presented in Figure 2.

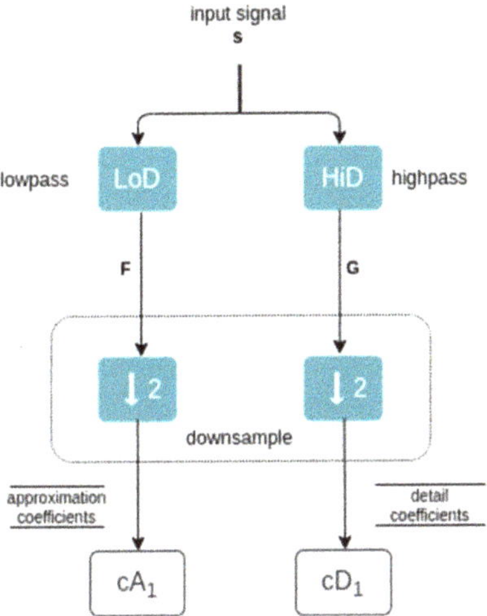

Figure 1. Wavelet decomposition algorithm.

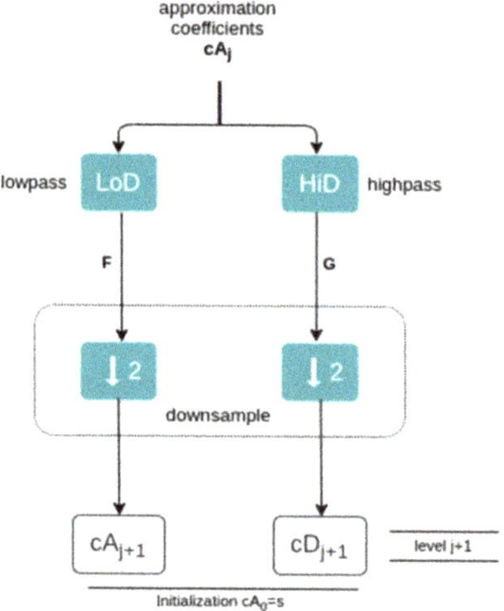

Figure 2. The 1D wavelet decomposition algorithm (wavedec function).

The decomposition was developed by Mallat [12]. DWT is commonly used for fast signal extraction [34]. The decomposition process is iterative, which means that the approximation component after iteration will be decomposed into several low-resolution components.

After the decomposition of the input signal s, we have the low-frequency coefficient cA1 and the high-frequency coefficient cD1 (see Figure 3). After the next decomposition of cA1, high-frequency component cD2 and low-frequency component cA2 are gained. If the decomposition process continues to level j = 3, in the last step coefficients cA3 and cD3 are obtained from cA2. Coefficient cA1 can be reconstructed by cA2 and cD2; similarly, cA2 can be reconstructed by cA3 and cD3.

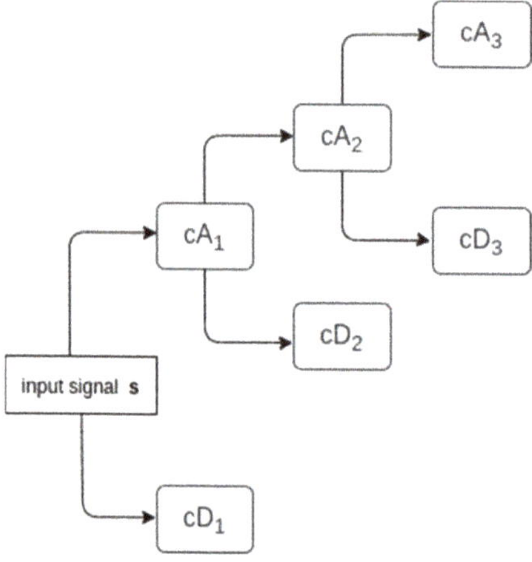

Figure 3. Three-level wavelet decomposition.

The wavelet transform is often used for data preparation for predictive systems with a neural network. In many cases, the type of neural network selected is very simple; oftentimes, a feedforward backpropagation neural network is used.

2.2. Artificial Neural Networks

An artificial neural network can be understood as a parallel processor that tends to preserve the knowledge gained through learning for its future use. They represent an artificial model of the human brain [53]. Knowledge is acquired by neurons during supervised learning, where the relationships between inputs and outputs are mapped. These relationships are created by neurons in the hidden layers that are interconnected. In the hidden layer, neurons capture and extract features from the inputs. The numbers of hidden layers and neurons may be different. Often, the number is selected by a trial-and-error approach, but optimization algorithms can also be used. Input-output relationships are regulated by weights, which are determined and adjusted during the learning process. The most common learning algorithm is backpropagation (BP). BP consists of two phases: forward and backward. During the forward phase, the response to the inputs is calculated; in the backward phase, the error between the response of the network and desired outputs is calculated. The calculated error is used for adjusting weights between inputs and outputs [54].

The efficiency of the three training algorithms was investigated in this study.

2.2.1. Scaled Conjugate Gradient

The principle of the BP algorithm is based on the calculation and adjustment of the weights in the steepest direction, which is time-consuming. The scaled conjugate gradient (SCG) uses at each iteration a different search direction, instead of the steepest direction that results in the fastest convergence. In the SCG algorithm, the line search technique, which is used to detect the step size, is omitted and replaced by quadratic approximation of the error function (Hessian matrix) together with the trust region from LM. This algorithm requires more iterations to converge but fewer computations between iterations [55].

2.2.2. Levenberg-Marquardt

The Levenberg-Marquardt (LM) algorithm was proposed to approach the second-order training speed without computing the Hessian matrix. The LM algorithm offers a tradeoff between the benefits of the Gauss-Newton method and the steepest descent method. To compute the connection weights wk, the LM uses the following Hessian matrix approximation:

$$w_{k+1} = w_k - \left[J^T + \mu I \right]^{-1} J^T e \tag{5}$$

where J is the Jacobian matrix (first-order derivatives of the errors), I is the unit matrix, e is the vector of network errors, and μ is the scalar parameter. Scalar parameter μ controls the algorithm. If $\mu = 0$, then the algorithm behaves as in Gauss-Newton's method. For a high value of μ, the algorithm uses the steepest descent method [56].

2.2.3. BFGS Quasi-Newton Backpropagation

The BFGS algorithm was proposed by Broyden, Fletcher, Goldfarb, and Shanno. The principle of Newton's method is based on computing the Hessian matrix. The BFGS algorithm is also based on Newton's method but uses the approximation of the Hessian matrix; then connection weights w_k are updated via the following Equation:

$$w_{k+1} = w_k - H^{-1} g \tag{6}$$

where H is the Hessian matrix (second-order derivatives of the errors) and g is the gradient. The BFGS algorithm is computationally more difficult and requires more storage due to computing and storing the approximation of the Hessian matrix [57].

3. Dataset Overview

The input data were provided by the local heating plant, which has more than 300 substations. For our experiments, we used historical load data from 1 November 2014 to 31

March 2015; for that period, we have a mixture of different data from various types of weather (see Figure 4). This dataset is used for training and testing. The total load consists of domestic hot water and the hot water used for central heating. The samples were logged every 10 min. Weather data were also collected from this heating plant. The units for load are in MW and for temperature in °C. Data were downloaded from SCADA in *.xlsx format, and then the data were converted into a suitable format for further processing. A large heat load drop at the beginning of the series was caused by the beginning of the heating season. The statistical characteristics of the data used are presented in Table 1.

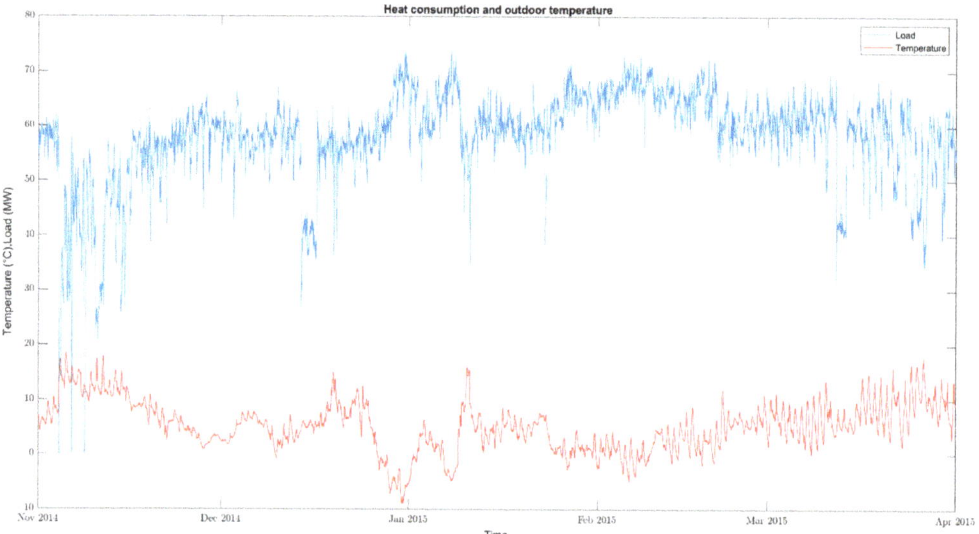

Figure 4. Heat consumption and outdoor temperature.

Table 1. Statistical characteristics of raw data.

Parameter	Min	Max	Mean	Std
Temperature (°C)	−8.93	18.47	5.27	4.52
Load (MW)	0	74.04	32.34	8.92

The scatter plot shown in Figure 5 shows the temperature and load dependency, as well as possible outliers. The presence of outliers can be explained by the fact that, during warmer days, the studied heat plant did not use the maximum power. That means that, during warmer periods, plants with less power were used.

Data preprocessing is necessary because all these data come from an industrial process, which is often full of errors such as missing data, noisy data, offset, etc. Missing data were substituted by the average of P_{-1} and P_{+1} values. Data normalization is a necessary step to avoid node saturation, which could negatively affect the training phase. For data normalization, we used Z score normalization:

$$Y_i = \frac{X_i - \overline{X}}{\sigma} \tag{7}$$

where Y_i is the normalized value, X_i is the actual value, $\overline{X}$ is the arithmetic mean, and σ is the standard deviation.

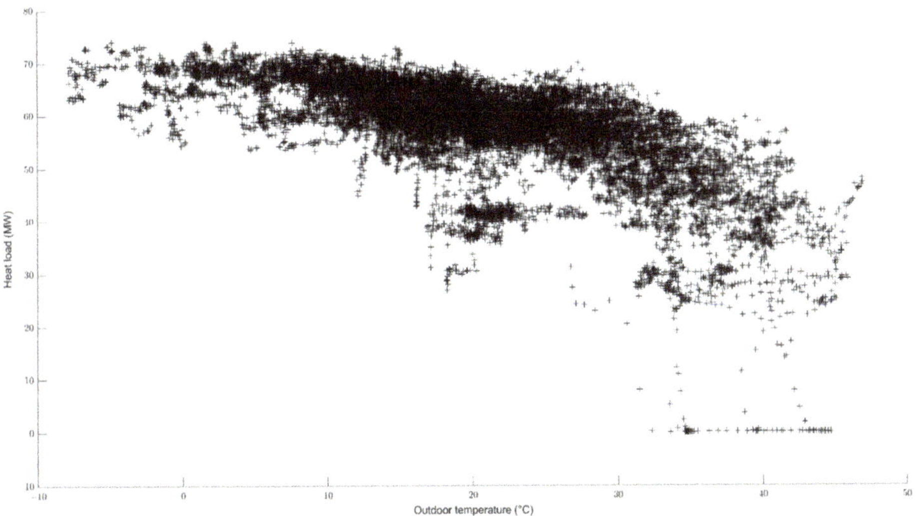

Figure 5. Scatter plot of temperature vs. load.

4. ANN and WANN Modeling

4.1. Mother Wavelet Selection Criteria

In the past, we can find many papers in the field of heat demand prediction where various researchers applied wavelet transform during the data preprocessing phase to predict heat consumption. The drawback of these studies is that the researchers chose mother wavelet empirically db4 [58], morlet [59]. There is an unwritten rule that Daubechies (db) mother wavelets, especially low order db2–db4, are the most suitable for load forecasting [60,61]. However, each signal has a unique characteristic and therefore we cannot rely on empirical selection. In this section, we are dealing with quantitative methods to select the appropriate mother wavelet. The investigated qualitative methods have different criteria for determining the appropriate mother wavelet; therefore, it is necessary to make a trade-off between criteria.

1. Maximum Energy Criteria

The base wavelet that maximizes energy from the wavelet coefficients represents the most appropriate wavelet for the analyzed signal. For a more detailed explanation, see [62].

$$E_{energy}(s) = \sum_{i=1}^{N} |wt(s,i)|^2 \tag{8}$$

where N is the number of wavelet coefficients and $wt(s,i)$ corresponds to the wavelet coefficients.

2. Minimum Shannon Entropy

The base wavelet that minimizes entropy from the wavelet coefficients represents the most appropriate wavelet for the analyzed signal. For a more detailed explanation, see [62].

$$E_{entropy}(s) = -\sum_{i=1}^{N} p_i \cdot \log_2 p_i \tag{9}$$

where p_i is the energy probability distribution of the wavelet coefficients, defined as

$$p_i = \frac{|wt(s,i)|^2}{E_{energy}(s)} \tag{10}$$

3.   Energy-to-Shannon Entropy ratio

The base wavelet that has produced the maximum energy-to-Shannon entropy ratio was selected to be the most appropriate wavelet for the analyzed signal [62]:

$$R(s) = \frac{E_{energy(s)}}{E_{entropy}(s)} \qquad (11)$$

where E_{energy} and $E_{entropy}$ are calculated using (Equations (8) and (9)).

In this section we compared 25 different wavelets from three different wavelet families, namely Daubechies, Symlets, and Coiflets (Figure 6). We also included the haar wavelet, which is often marked as db1. The best wavelet was selected based on the criteria mentioned above. For the first criterion, maximum energy, the best wavelet was db1 or haar. Based on the second criterion, minimum Entropy, the most suitable wavelet was sym 10. This conclusion contradicts the first criterion. To avoid such a conflict, it is necessary to find a trade-off between criteria. Based on the maximum energy-entropy ratio, the db5 wavelet produced the highest value, which means that that db5 wavelet is the optimal mother wavelet for our purposes.

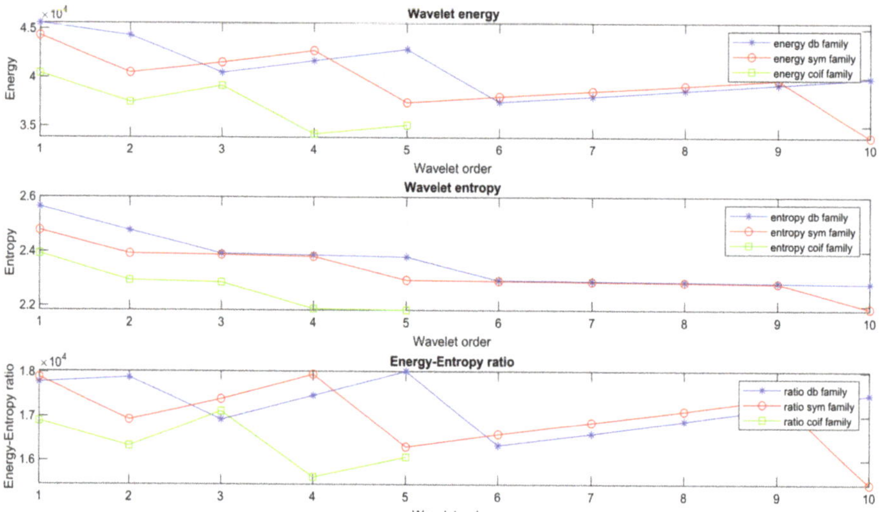

Figure 6. Mother wavelet selection—quantitative criteria.

4.2. Decomposition Level Selection

A suitable decomposition level is as important as selecting the mother wavelet. A high number of decomposition levels might cause information loss, high computational effort, etc. [63]. In the past, several papers were published where researchers used the following empirical equation: $L = \log_2(N)$, where N is the series length, to determine the decomposition level [64]. In our case, it would be 14 levels. The difference between the raw heat load series and decomposed series is clearly visible after approximation at level 11. We set the maximum decomposition level to 11; this decomposition level also involves all possible mother wavelet candidates [65]:

$$L = \frac{\log\left(\frac{N}{lw-1}\right)}{\log 2} \qquad (12)$$

where N is the series length and lw is the wavelet filter size.

We focused on a deeper analysis of the decomposed signal via the proposed method by Sang [66]. This method is used to identify the true and noisy components of each decomposition level. The comparison of the energy of raw series and referenced noise series identified components that are close to or inside of the confidence interval of the referenced noise series. In Figure 7, we can clearly identify the components (D1–D3) that are likely to be noise. We suppose that the components above level D4 are true components of the signal. From this analysis, we propose two suitable decompositions at level 6 and level 9 by db5 mother wavelet.

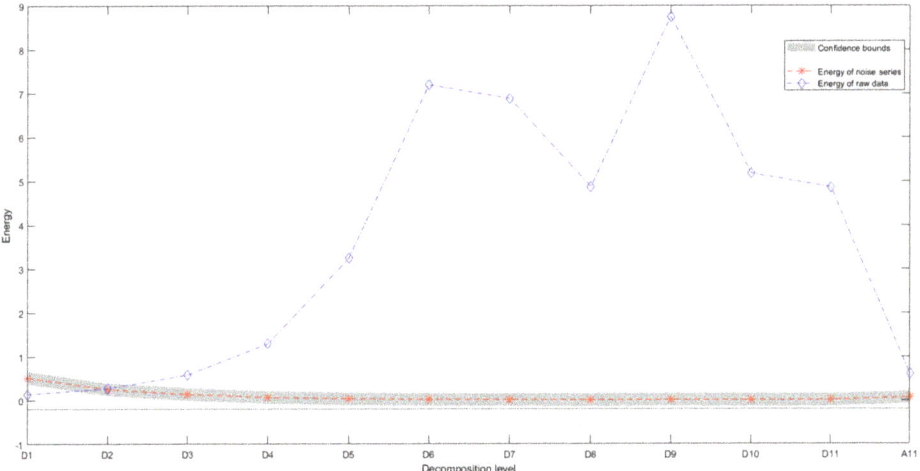

Figure 7. Energy of subseries.

4.3. Building WANN and ANN Models

For an accurate forecast of heat consumption, it is necessary to make a detailed analysis of the variables that influence heat consumption. Generally, energy consumption depends on many factors such as social and climate parameters, type of consumers, etc. Heat demand strongly depends on the outdoor temperature and other climate factors like humidity, wind speed, and so on. Research papers from the past confirm that the strongest influence on demand is outdoor temperature [67]. That fact is also proven by a correlation analysis between heat load and outdoor temperature, where the correlation coefficient is -0.78, which could be considered a strong relationship. Another good predictor is historical load. It is likely that consumption the following day at the same time will be similar to the consumption the day before. Significant lags were determined by autocorrelation analysis. Generally, heat load consumption also depends on the time of the day and day of the week. These factors could also increase the prediction accuracy. Table 2 shows the selected input variables for WANN.

Table 2. Selected input variables.

Input Number	Input Name	Value	Calculation
1.	Hour	0–23	
2.	Weekend	0–1	Timestamp
3.	Day of the week	1–7	
4.	Temperature	Various	Exogenous
5.	Lagged load	Various	Endogenous + timestamp

1. Hour—To capture the cyclical behavior of the series, the hour variable was encoded via sine and cosine transform:

$$Hsin = \frac{\sin(2\pi h)}{24} \tag{13}$$

$$Hcos = \frac{\cos(2\pi h)}{24} \tag{14}$$

Weekend—1 represents weekends and 0 represents weekdays.

Day of the week *(DoW)*—determines the days of the week, where Mondays are marked as 1 and Sundays as 7.

$$DOWsin = \frac{\sin(2\pi dow)}{7} \tag{15}$$

$$DOWcos = \frac{\cos(2\pi dow)}{7} \tag{16}$$

Temperature $T(t)$—The temperature values at lags t, t_{-144}. Because of small differences in temperature, the average temperature at t_{-1} and t_{-2} is used. Lagged load $L(t)$—Autocorrelation analysis was used to select the most relevant historical consumption. A window of length 1008 (one week) was considered for selection. Selected lags are listed in Table 3.

Table 3. Selected lags and proposed WANN model structure.

| Number | Selected Lags | | Model Structure |
	L(t)	T(t)	(I $\times$ h $\times$ o)
D1	1–5	–	10 $\times$ h $\times$ 1
D2	1–4, 6	–	10 $\times$ h $\times$ 1
D3	1,2,4,5,6	–	10 $\times$ h $\times$ 1
D4	1–3,10–12	–	11 $\times$ h $\times$ 1
D5	1–4,20–24	–	14 $\times$ h $\times$ 1
D6	1–5, 39–42	✓	18 $\times$ h $\times$ 1
D7	1–5, 79–81	–	13 $\times$ h $\times$ 1
D8	1–5, 172–174	–	13 $\times$ h $\times$ 1
D9	1–5, 319–321	✓	17 $\times$ h $\times$ 1
A9	1–5	✓	14 $\times$ h $\times$ 1
A6	1–5	✓	14 $\times$ h $\times$ 1

Determining a suitable number of hidden neurons and hidden layers is crucial in ANN modeling. Too many hidden neurons could cause overfitting. To avoid overfitting, it is crucial to select an appropriate number of neurons in the hidden layer h. Unfortunately, there is no equation to compute the number of hidden neurons. In most cases, the appropriate number of hidden neurons is determined by trial and error. In our research, the number of hidden neurons was determined by rule of thumb: the number of hidden neurons is two-thirds the size of the input layer [67]. The parameters of the proposed models are presented in Table 4.

Table 4. Parameters of proposed feedforward neural networks.

Model	Parameter	Value
BPNN and WANN	Number of hidden layers	1
	Number of neurons in hidden layer	21 for ANN models Various for WANN; see Table 3
	Number of output neurons	1
	Hidden layer activation function	tansig
	Output layer activation function	purelin
	Data set division train/test	random 80/20 (%)
	Epochs	1000
	Data normalization	mapstd; see (Equation (7))
	Training algorithms	trainlm, trainscg, trainbfg
	Learning rate	0.001
WANN	Decomposition level	6 and 9
	Mother wavelet	db5

The proposed methodology is depicted in Figure 8 and consists of the following steps:

1. Load input raw data;
2. Decompose load data using DWT into N subseries of details and approximations;
3. Perform feature selection—autocorrelation, correlation analysis;
4. Normalize data using mapstd function;
5. Create an input matrix from selected features;
6. Divide the processed data into training and testing sets;
7. Create WANN models
8. Compute the number of hidden neurons (2/3 of inputs)
9. Train and test until error starts to increase, then stop training;
10. Reconstruct predicted outputs and reconstruct signal $X_{rec} = D_{1+}, \ldots, + D_n + A_n$;
11. Denormalize outputs using reverse mapstd function;
12. Validate proposed models on a new dataset.

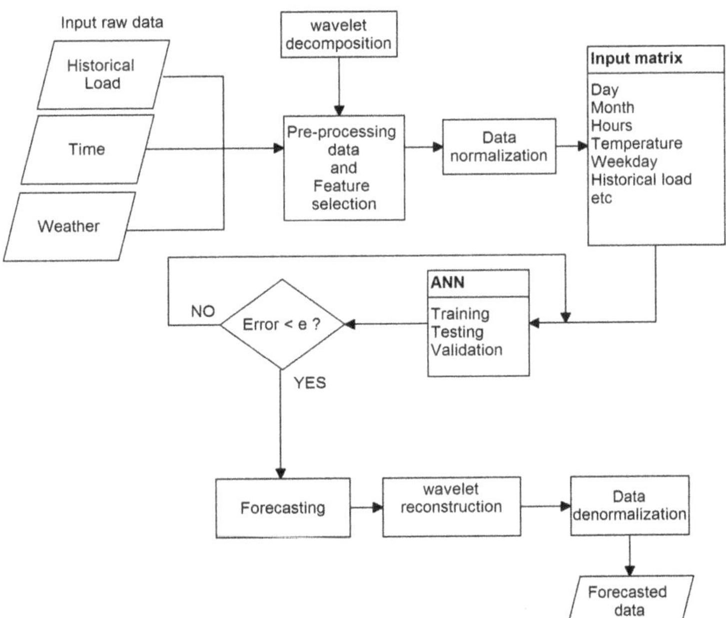

Figure 8. Proposed WANN model flowchart.

5. Results and Discussion

The results of the proposed models for 1 h ahead (10 min sampling interval) are presented in this section. In this research, three FFBP ANNs were proposed and three different learning algorithms, Levenberg-Marquardt (LM), BFGS quasi-Newton, and Scaled Conjugate Gradient (SCG), were tested. The ANN models were compared with WANNs. The experiments were performed in the MATLAB2020a environment on a laptop with an i7 3.00 GHz CPU and 16 GB memory.

5.1. Evaluation Metrics

To evaluate the accuracy of the proposed models, the following metrics were used: mean absolute error (MAE), mean absolute percentage error (MAPE), and root mean square error (RMSE). The definition of these metrics is as follows:

$$\text{MAE} = \frac{1}{n}\sum_{i=1}^{n}\left|Y_i - \overline{Y_i}\right| \tag{17}$$

$$\text{MAPE} = \frac{1}{n}\sum_{i=1}^{n}\left|\frac{Y_i - \overline{Y_i}}{Y_i}\right| \tag{18}$$

$$\text{RMSE} = \sqrt{\frac{1}{n}\sum_{i=1}^{n}\left(Y_i - \overline{Y_i}\right)^2} \tag{19}$$

where n—number of samples, Y_i—actual value, $\overline{Y_i}$—predicted value. The smaller results represent better prediction accuracy.

5.2. WANN and ANN Prediction Comparison

The accuracy of the proposed WANN and ANN models was validated on a dataset from February 2018. This dataset was not included in the training and testing models. Figures 9 and 10 shows the prediction error of the decomposed series with different learning algorithms. Upon visual checking, it is noticeable that the prediction accuracy for the wavelet details D1, D2, and D3 shows high differences compared to other details and there are significant errors according to the other details. The MAPE was 322.93% for LM, and SCG and BFG produced MAPE over 370%. This is caused by high-frequency components (i.e., noise), but these sub-bands also contain some useful features that are predictable. As stated in Section 4.2, sub-bands D1-D3 contain noise and could be omitted. The error rate rapidly decreases after D4. The presence of a higher error rate is also clearly visible in the A9 series, where BFG and SCG produce much worse predictions compared to the LM algorithm. Both algorithms have a tendency to overestimate the load with the total MAPE for the SCG (0.135%), 0.112% (BFG), and 0.017% (LM). In general, the WANN model shows a good ability to capture the features from the decomposed raw data.

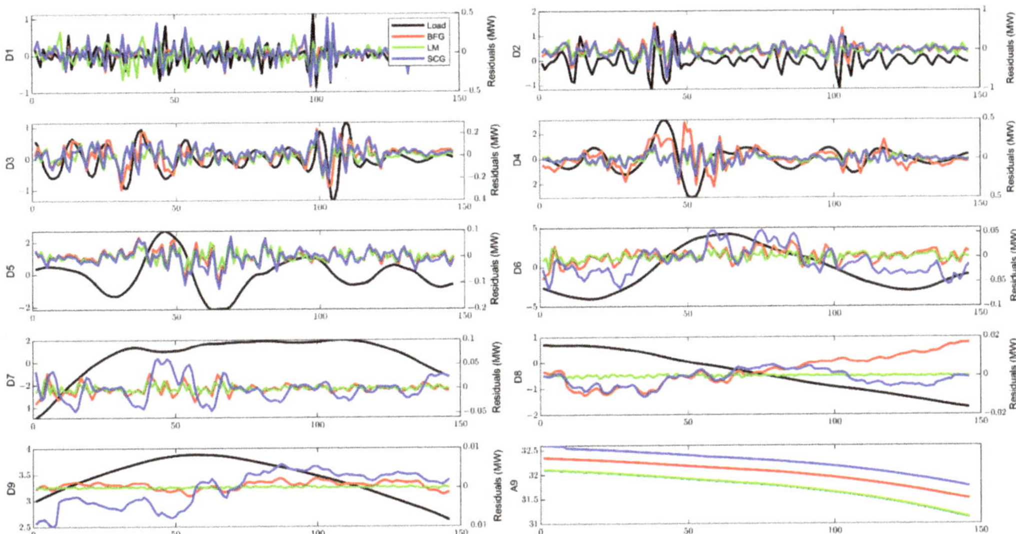

Figure 9. Residuals vs. real load of decomposed series.

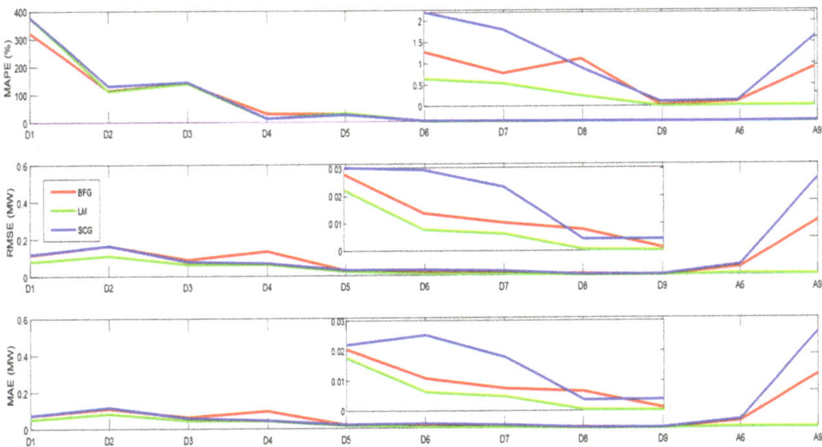

Figure 10. MAPE, RMSE, and MAE performance of the training algorithms.

The error analysis from Figure 10 indicates that the suitable decomposition level is 6. Predicted sub-bands *D1, Dn, An* were reconstructed to obtain the total heat consumption, which is presented in Figure 11.

The accuracy of the WANNs' prediction is better compared to conventional ANN. In every experiment, WANN models outperform the conventional ANN model. In each case, the LM algorithm produced the highest accuracy. The MAPE for conventional ANN LM was 1.75%, for WANN LM6 it was 0.359%, and for model WANN LM9 it was 0.362%. The difference between the WANN LM6 and WANN LM9 inaccuracy is negligible. Other values are presented in Table 5 and Figure 12.

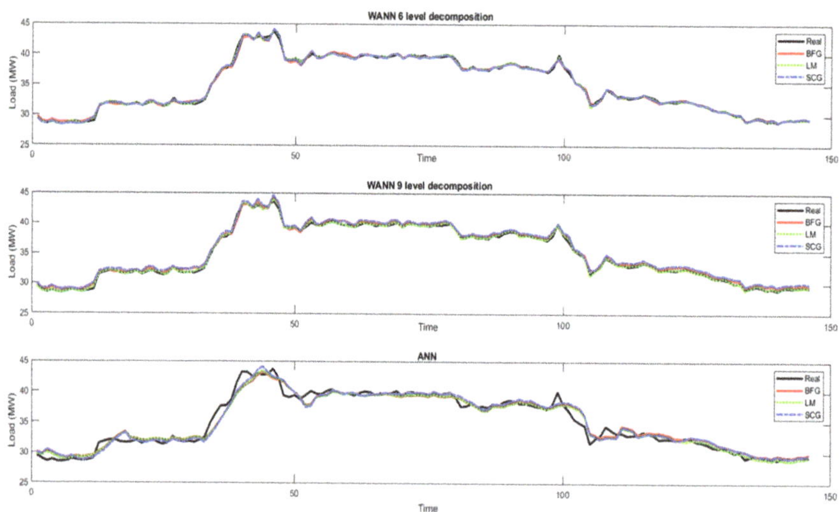

Figure 11. Comparison of reconstructed WANN and ANN (1 h ahead).

Table 5. Errors obtained by ANN and WANN models.

Models	ANN			WANN					
				Dec. Level 6			Dec. Level 9		
Parameters	BFG	LM	SCG	BFG	LM	SCG	BFG	LM	SCG
MAPE (%)	1.91	1.75	1.83	0.58	0.36	0.51	0.98	0.36	1.51
RMSE (MW)	0.88	0.85	0.89	0.28	0.16	0.25	0.39	0.16	0.56
MAE (MW)	0.66	0.61	0.64	0.20	0.12	0.18	0.33	0.12	0.51

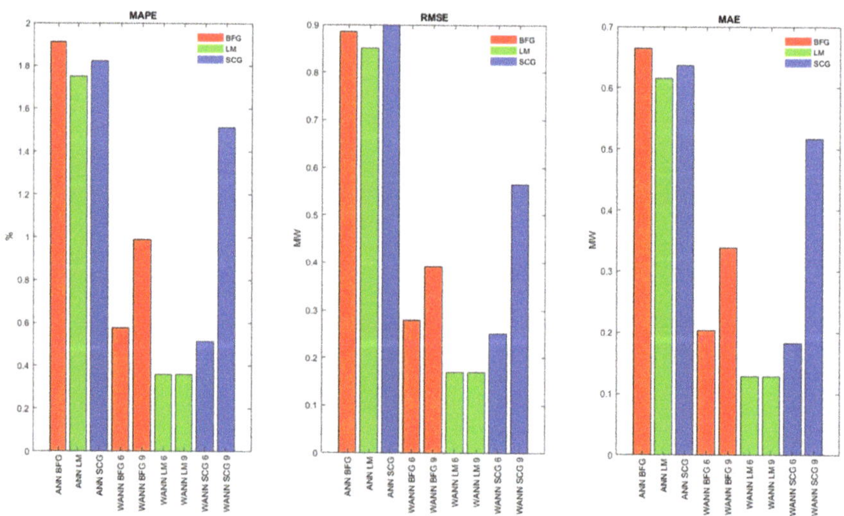

Figure 12. Prediction comparison of different algorithms.

It is worth mentioning that the proposed WANN models in each configuration were able to improve accuracy in every aspect. Percentage improvements are listed in Table 6, which demonstrates the importance of employing DWT in the data preprocessing stage.

Table 6. WANN improvement percentage over conventional ANN.

| | WANN | | | | | |
| Improvement percentage | Dec. level 6 | | | Dec. level 9 | | |
	BFG	LM	SCG	BFG	LM	SCG
MAPE	69%	**79%**	72%	48%	**79%**	17%
RMSE	68%	**81%**	71%	55%	**81%**	37%
MAE	**81%**	80%	71%	50%	**80%**	20%

Table 6 shows that the improvement had a decreasing tendency after decomposition level 6 for the BFG and SCG algorithms. The largest percentage decrease in accuracy was produced by the SCG algorithm in each evaluation metric. The drop was 51% for MAE. However, these decreases in accuracy represented higher overall accuracy compared to conventional ANNs. The LM algorithm produced the biggest improvement in every metric. Compared to the conventional model, the increase was 79%. The LM algorithm showed no improvement after decomposition level 6.

6. Conclusions

This research paper dealt with the prediction of heat consumption. Several models have been created that can predict heat consumption with varying accuracy. Some important findings have been identified during this research. The mother wavelet was chosen based on a quantitative criterion, the energy-entropy ratio. According to Figure 6, the most suitable mother wavelet was db5. From the presented results, it is clear that the suitable decomposition level is 6. The accuracy of the reconstructed signals shows that models with decomposition level 6 have better results compared to models with decomposition level 9 (Figures 10 and 11). Also, we can state that some details could be omitted during the signal reconstruction. Several models were created for the purpose of finding the most appropriate training algorithm. The presented results show that models trained with an LM algorithm outperform other models (Table 5). Calculation of the error metrics MAPE, RMSE, and MAE proved that the LM training algorithm offered the best results for all models. This research also compared the effectiveness of employing wavelet transform during data preprocessing. In each case, the WANN models predicted heat consumption with significantly higher accuracy compared to ANNs. Significant differences in accuracy were achieved in every WANN model. The LM algorithm produced the highest accuracy among WANN and ANN models. Compared to the conventional model (1.75% MAPE), the improvement was near five times greater (0.36% MAPE). The highest error was produced by the BFG algorithm in both cases. We can state that a combination of wavelet decomposition and ANN could significantly improve the prediction performance. The outcome of this research is also a MATLAB GUI application that could be used by dispatchers. In future research, there are several opportunities to improve the models, e.g., propose and test other ANN architectures like Elman, RNN, optimize the number of hidden neurons with PSO, reduce input parameters with PCA, and propose models with a longer forecasting period (12 h ahead, 24 h ahead).

Author Contributions: Conceptualization, G.M. and P.V.; methodology, S.K.; software, S.K.; validation, S.K., G.M., and I.H.; formal analysis, I.H.; investigation, I.H.; resources, I.H.; data curation, S.K.; writing—original draft preparation, S.K.; writing—review and editing, I.H.; visualization, S.K.; supervision, G.M.; project administration, P.V.; funding acquisition, P.V. All authors have read and agreed to the published version of the manuscript.

Funding: This research was funded by Mladý výskumník, "Návrh neurónovej siete na predikciu spotreby tepla"; the Scientific Grant Agency of the Ministry of Education, Science, Research and Sport of the Slovak Republic and the Slovak Academy of Sciences, grant number VEGA 1/0272/18, "Holistic approach of knowledge discovery from production data in compliance with Industry 4.0 concept"; and the Scientific Grant Agency of the Ministry of Education, Science, Research and Sport of the Slovak Republic and the Slovak Academy of Sciences, grant number 1/0232/18, "Using the methods of multiobjective optimization in production processes control."

Acknowledgments: We would like to thank BAT company for providing heat consumption data and the anonymous reviewers for their comments, which improved the quality of the work. This publication is the results of the project ITMS 313011W988: "Research in the SANET network and possibilities of its further use and development" within the Operational Program Integrated Infrastructure co-financed by the ERDF.

Conflicts of Interest: The authors declare no conflict of interest.

References

1. Gabriska, D. Evaluation of the Level of Reliability in Hazardous Technological Processes. *Appl. Sci.* **2021**, *11*, 134. [CrossRef]
2. Kurek, T.; Bielecki, A.; Świrski, K.; Wojdan, K.; Guzek, M.; Białek, J.; Brzozowski, R.; Serafin, R. Heat Demand forecasting algorithm for a Warsaw district heating network. *Energy* **2021**, *217*. [CrossRef]
3. Guo, B.; Cheng, L.; Xu, J.; Chen, L. Prediction of the Heat Load in Central Heating Systems Using GA-BP Algorithm. In Proceedings of the International Conference on Computer Network, Electronic and Automation (ICCNEA 2017), Xi'an, China, 23–25 September 2017; pp. 441–445. [CrossRef]
4. GRANRYD Eric. *Refrigerating engineering*; Royal Institute of Technology: Stockholm, Sweden, 2009; ISBN 978-91-7415-415-3.
5. Panapakidis, I.P.; Dagoumas, A.S. Day-ahead natural gas demand forecasting based on the combination of wavelet transform and ANFIS/genetic algorithm/neural network model. *Energy* **2017**, *118*, 231–245. [CrossRef]
6. Yan, K.; Li, W.; Ji, Z.; Du, Y.; Qi, M. A Hybrid LSTM Neural Network for Energy Consumption Forecasting of Individual Households. *IEEE Access* **2019**, *7*, 157633–157642. [CrossRef]
7. Nemeth, M.; Borkin, D.; Michalconok, G. The comparison of machine-learning methods XGBoost and LightGBM to predict energy development. In Proceedings of the Computational Statistics and Mathematical Modeling Methods in Intelligent Systems: Proceedings of 3rd Computational Methods in Systems and Software, Zlín, Czech Republic, 10–12 September 2019; Silhavy, R., Silhavy, P., Prokopova, Z., Eds.; Springer: Cham/Basel Switzerland, 2019; Volume 2, pp. 208–215. [CrossRef]
8. Nemetova, A.; Borkin, D.; Michalconok, G. Comparison of methods for time series data analysis for further use of machine learning algorithms. In Proceedings of the Computational Statistics and Mathematical Modeling Methods in Intelligent Systems: Proceedings of 3rd Computational Methods in Systems and Software, Zlín, Czech Republic, 10–12 September 2019; Silhavy, R., Silhavy, P., Prokopova, Z., Eds.; Springer: Cham/Basel Switzerland, 2019; Volume 2, pp. 90–99. [CrossRef]
9. Lang, X.; Rehman, N.; Zhang, Y.; Xie, L.; Su, H. Median ensemble empirical mode decomposition. *Signal Process.* **2020**, *176*. [CrossRef]
10. Zuo, G.; Luo, J.; Wang, N.; Lian, Y.; He, X. Decomposition ensemble model based on variational mode decomposition and long short-term memory for streamflow forecasting. *J. Hydrol.* **2020**, *585*. [CrossRef]
11. Yesilli, M.C.; Khasawneh, F.A.; Otto, A. On transfer learning for chatter detection in turning using wavelet packet transform and ensemble empirical mode decomposition. *Cirp J. Manuf. Sci. Technol.* **2020**, *28*, 118–135. [CrossRef]
12. Mallat, S.G. *A Theory for Multiresolution Signal Decomposition: The Wavelet Representation*; Technical Report; University of Pennsylvania: Philadelphia, PA, USA, 1987.
13. Meyer, Y. *Wavelets, Algorithms & Applications*, 1st ed.; SIAM: Philadelphia, PA, USA, 1993.
14. Sui, K.; Kim, H.G. Research on application of multimedia image processing technology based on wavelet transform. *J. Image Video Process.* **2019**, *24*. [CrossRef]
15. Mahesh, M.; Kumar, T.R.R.; Shoban Babu, B.; Saikrishna, J. Image Enhancement using Wavelet Fusion for Medical Image Processing. *Int. J. Eng. Adv. Technol.* **2019**, *9*. [CrossRef]
16. Shanmugapriya, K.; Priya, D.J.; Priya, N. Image Enhancement Techniques in Digital Image Processing. *Int. J. Innov. Technol. Explor. Eng.* **2019**, *8*. [CrossRef]
17. Kumar, K.; Mustafa, N.; Li, J.; Shaikh, R.A.; Khan, S.A.; Khan, A. Image edge detection scheme using wavelet transform. In Proceedings of the International Computer Conference on Wavelet Active Media Technology and Information Processing (ICCWAMTIP 2014), Chengdu, China, 19–21 December 2014; pp. 261–265. [CrossRef]
18. Hashim, M.A.; Nasef, M.H.; Kabeel, A.E.; Ghazaly, N.M. Combustion fault detection technique of sparkignition engine based on wavelet packet transform and artificial neural network. *Alex. Eng. J.* **2020**. [CrossRef]
19. Gharesi, N.; Mehdi Arefi, M.; Razavi-Farb, R.; Zarei, J.; Yin, S. A neuro-wavelet based approach for diagnosing bearing defects. *Adv. Eng. Inform.* **2020**, *46*. [CrossRef]
20. Kou, L.; Liu, C.; Cai, G.; Zhang, Z. Fault Diagnosis for Power Electronics Converters based on Deep Feedforward Network and Wavelet Compression. *Electr. Power Syst. Res.* **2020**, *185*. [CrossRef]

21. Valizadeh, M.; Sohrabi, M.R.; Motiee, F. The application of continuous wavelet transform based on spectrophotometric method and high-performance liquid chromatography for simultaneous determination ofanti-glaucoma drugs in eye drop. *Spectrochim. Acta Part A Mol. Biomol. Spectrosc.* **2020**, *242*. [CrossRef]
22. Győrfi, Á.; Szilágyi, L.; Kovács, L. A Fully Automatic Procedure for Brain Tumor Segmentation from Multi-Spectral MRI Records Using Ensemble Learning and Atlas-Based Data Enhancement. *Appl. Sci.* **2021**, *11*, 564. [CrossRef]
23. Akansu, A.N.; Serdijn, W.A.; Selesnick, W.I. Emerging applications of wavelets: A review. *Phys. Commun.* **2010**, *3*. [CrossRef]
24. Zuo, H.; Chen, Y.; Jia, F. A new C0 layer wise wavelet finite element formulation for the static and free vibration analysis of composite plates. *Compos. Struct.* **2020**, *254*. [CrossRef]
25. Qin, Y.; Mao, Y.; Tang, B.; Wang, Y.; Chen, H. M-band flexible wavelet transform and its application to thefault diagnosis of planetary gear transmission systems. *Mech. Syst. Signal Process.* **2019**, *134*. [CrossRef]
26. Zhao, Z.; Wang, X.; Zhang, Y.; Gou, H.; Yang, F. Wind speed prediction based on wavelet analysis and time series method. In Proceedings of the International Conference on Wavelet Analysis and Pattern Recognition (ICWAPR 2017), Ningbo, China, 9–12 July 2017; pp. 23–27. [CrossRef]
27. Ren, P.; Xiang, Z.; Shangguan, R. Design and Simulation of a prediction algorithm based on wavelet support vector machine. In Proceedings of the Seventh International Conference on Natural Computation, Shanghai, China, 26–28 July 2011; pp. 208–211. [CrossRef]
28. Barthel, K.U.; Brandau, S.; Hermesmeier, W.; Heising, G. Zerotree wavelet coding using fractal prediction. In Proceedings of the International Conference on Image Processing (ICIP 1997), Santa Barbara, CA, USA, 26–29 October 1997; Volume 2, pp. 314–317. [CrossRef]
29. Yin, J.; Gao, C.; Wang, Y.; Wang, Y. Hyperspectral image classification using wavelet packet analysis and gray prediction model. In Proceedings of the International Conference on Image Analysis and Signal Processing (IASP 2010), Zhejiang, China, 9–11 April 2010; pp. 322–326. [CrossRef]
30. Elarabi, T.; Sammoud, A.; Abdelgawad, A.; Li, X.; Bayoumi, M. Hybrid wavelet—DCT intra prediction for H.264/AVC interactive encoder. In Proceedings of the IEEE China Summit & International Conference on Signal and Information Processing (ChinaSIP 2014), Xi'an, China, 9–13 July 2014; pp. 281–285. [CrossRef]
31. Li, H.; Guo, C.; Yang, S.X.; Jin, H. Hybrid Model of WT and ANFIS and Its Application on Time Series Prediction of Ship Roll Motion. In Proceedings of the Multiconference on Computational Engineering in Systems Applications (CESA 2006), Beijing, China, 4–6 October 2006; pp. 333–337. [CrossRef]
32. Stefenon, S.F.; Ribeiro, M.H.D.M.; Nied, A.; Mariani, V.C.; Coelho, L.D.S.; da Rocha, D.F.M.; Grebogi, R.B.; Ruano, A.E.D.B. Wavelet group method of data handling for fault prediction in electrical power insulators. *Int. J. Electr. Power Energy Syst.* **2020**, *123*. [CrossRef]
33. Prabhakar, D.V.N.; Kumar, M.S.; Krishna, A.G. A Novel Hybrid Transform approach with integration of Fast Fourier, Discrete Wavelet and Discrete Shearlet Transforms for prediction of surface roughness on machined surfaces. *Measurement* **2020**, *164*. [CrossRef]
34. Zhang, T.; Chaofeng, L.; Fumin, M.; Zhao, K.; Wang, H.; O'Hare, G.M. A photovoltaic power forecasting model based on dendritic neuron networks with the aid of wavelet transform. *Neurocomputing* **2020**, *397*, 438–446. [CrossRef]
35. Chang, Z.; Zhang, Y.; Chen, W. Electricity price prediction based on hybrid model of Adam optimized LSTM neural network and wavelet transform. *Energy* **2019**, *187*. [CrossRef]
36. El-Hendawi, M.; Wang, Z. An ensemble method of full wavelet packet transform and neural network for short term electrical load forecasting. *Electr. Power Syst. Res.* **2020**, *182*. [CrossRef]
37. Liu, H.; Mi, X.; Li, Y. Smart deep learning based wind speed prediction model using wavelet packet decomposition, convolutional neural network and convolutional long short term memory network. *Energy Convers. Manag.* **2018**, *166*, 120–131. [CrossRef]
38. Xia, C.; Zhang, M.; Cao, J. A hybrid application of soft computing methods with wavelet SVM and neural network to electric power load forecasting. *J. Electr. Syst. Inf. Technol.* **2018**, *5*, 681–696. [CrossRef]
39. Bashir Tayab, U.; Zia, A.; Yang, F.; Lu, J.; Kashif, M. Short-term load forecasting for microgrid energy management system using hybrid HHO-FNN model with best-basis stationary wavelet packet transform. *Energy* **2020**, *203*. [CrossRef]
40. Farhadi, M.; Abbaspour-Gilandeh, Y.; Mahmoudi, A.; Mari Maja, J. An Integrated System of Artificial Intelligence and Signal Processing Techniques for the Sorting and Grading of Nuts. *Appl. Sci.* **2020**, *10*, 3315. [CrossRef]
41. Feng, X.; Feng, Q.; Li, S.; Hou, X.; Zhang, M.; Liu, S. Wavelet-Based Kalman Smoothing Method for Uncertain Parameters Processing: Applications in Oil Well-Testing Data Denoising and Prediction. *Sensors* **2020**, *20*, 4541. [CrossRef]
42. Obidin, M.V.; Serebrovski, A.P. Signal denoising with the use of the wavelet transform and the Kalman filter. *J. Commun. Technol. Electron.* **2014**, *59*, 1440–1445. [CrossRef]
43. Li, Y.J.; Kokkinaki, A.; Darve, E.T.; Kitanidis, P.K. Smoothing-based compressed state Kalman filter for joint state-parameter estimation: Applications in reservoir characterization and CO_2 storage monitoring. *Water Resour. Res.* **2017**, *53*, 7190–7207. [CrossRef]
44. Zhang, X.; Ni, W.; Liao, H.; Pohl, E.; Xu, P.; Zhang, W. Fusing moving average model and stationary wavelet decomposition for automatic incident detection: Case study of Tokyo Expressway. *J. Traffic Transp. Eng.* **2014**, *1*, 404–414. [CrossRef]
45. Szi-Wen, C.; Hsiao-Chen, C.; Hsiao-Lung, C. A real-time QRS detection method based on moving-averaging incorporating with wavelet denoising. *Comput. Methods Programs Biomed.* **2006**, *82*, 187–195.

46. Akansu, A.N.; Haddad, R.A. *Multiresolution Signal Decomposition: Transforms, Subbands, and Wavelets*, 2nd ed.; Academic Press: San Diego, CA, USA, 2000.
47. Tan, L.; Jiang, J. Discrete Wavelet Transform. In *Digital Signal Processing—Fundamentals and Applications*, 3rd ed.; Elsevier: Amsterdam, The Netherlands, 2019; pp. 623–632.
48. Boashash, B. The Discrete Wavelet Transform. In *Time-Frequency Signal Analysis and Processing—A Comprehensive Reference*; Elsevier: Amsterdam, The Netherlands, 2016; pp. 141–142.
49. Loizou, C.P.; Pattichis, C.S.; D'hooge, J. Discrete Wavelet Transform. In *Handbook of Speckle Filtering and Tracking in Cardiovascular Ultrasound Imaging and Video*; Institution of Engineering and Technology: London, UK, 2018; pp. 174–177.
50. Bankman, I.N. Three-Dimensional Image Compression with Wavelet Transforms. In *Handbook of Medical Image Processing and Analysis*, 2nd ed.; Elsevier: Amsterdam, The Netherlands, 2009; pp. 963–964.
51. MathWorks Wavedec. Available online: https://www.mathworks.com/help/wavelet/ref/wavedec.html (accessed on 1 November 2020).
52. Freire, P.K.D.M.M.; Santos, C.A.G.; da Silva, G.B.L. Analysis of the use of discrete wavelet transforms coupled with ANN for short-term streamflow forecasting. *Appl. Soft Comput.* **2019**, *80*, 494–505. [CrossRef]
53. Junior, L.A.; Souza, R.M.; Menezes, M.L.; Cassiano, K.M.; Pessanha, J.F.; Souza, R. Artificial Neural Network and Wavelet Decomposition in the Forecast of Global Horizontal Solar Radiation. *Pesqui. Oper.* **2015**, *35*, 73–90. [CrossRef]
54. Moller, M.F. A scaled conjugate gradient algorithm for fast supervised learning. *Neural Netw.* **1993**, *6*, 525–533. [CrossRef]
55. Rodrigues, F.; Cardeira, C.; Calado, J.M.F. The Daily and Hourly Energy Consumption and Load Forecasting Using Artificial Neural Network Method: A case Study Using a Set of 93 Households in Portugal. *Energy Procedia* **2014**, *62*, 220–229. [CrossRef]
56. Perera, A.; Azamathulla, H.; Rathnayake, U. Comparison of different Artificial Neural Network (ANN) training algorithm to predict atmospheric temperature in Tabuk, Saudi Arabia. *Mausam* **2020**, *25*, 1–11.
57. Gong, M.; Wang, J.; Bai, Y.; Li, B.; Zhang, L. Heat load prediction of residential buildings based on discrete wavelet transform and tree-based ensemble learning. *J. Build. Eng.* **2020**, *32*. [CrossRef]
58. Wang, M.; Qi, T. Application of wavelet neural network on thermal load forecasting. *Int. J. Wirel. Mob. Comput.* **2013**, *6*, 608–614. [CrossRef]
59. Amjady, N.; Keynia, F. Short-term load forecasting of power systems by com- bination of wavelet transform and neuro-evolutionary algorithm. *Energy* **2009**, *34*, 46–57. [CrossRef]
60. Bashir, Z.A.; El-Hawary, M.E. Applying wavelets to short-term load forecasting using PSO-based neural networks. *IEEE Trans. Power Syst.* **2009**, *24*, 20–27. [CrossRef]
61. Gao, R.X.; Yan, R. Wavelets: Theory and applications for manufacturing. *Wavelets Theory Appl. Manuf.* **2011**, 165–187. [CrossRef]
62. Tascikaraoglu, A.; Sanandaji, B.M.; Poolla, K.; Varaiya, P. Exploiting sparsity of interconnections in spatio-temporal wind speed forecasting using Wavelet Transform. *Appl. Energy* **2016**, *165*, 735–747. [CrossRef]
63. Daubechies, I. *Ten Lectures on Wavelets, CBMS-NSF Regional Conference Series in Applied Mathematics*; SIAM: Philadelphia, PA, USA, 1992.
64. Freire, P.K.D.M.; Santos, C.A.G. Optimal level of wavelet decomposition for daily inflow forecasting. *Earth Sci. Inf.* **2020**, *13*, 1163–1173. [CrossRef]
65. Sang, Y. A Practical Guide to Discrete Wavelet Decomposition of Hydrologic Time Series. *Water Resour. Manag.* **2012**, *26*, 3345–3365. [CrossRef]
66. Yang, H.; Jin, S.; Feng, S.; Wang, B.; Zhang, F.; Che, J. Heat Load Forecasting of District Heating System Based on Numerical Weather Prediction Model. In Proceedings of the 2nd International Forum on electrical Engineering and Automation (IFEEA 2015), Guangzhou, China, 26–27 December 2015; pp. 1–5. [CrossRef]
67. Karsoliya, S.; Azad, M. Approximating Number of Hidden layer neurons in Multiple Hidden Layer BPNN Architecture. *Int. J. Eng. Trends Technol.* **2012**, *3*, 714–717.

Article

An Artificial Intelligence Empowered Cyber Physical Ecosystem for Energy Efficiency and Occupation Health and Safety

Petros Koutroumpinas [1], Yu Zhang [2,*], Steve Wallis [3] and Elizabeth Chang [2]

[1]   Faculty of Engineering, Monash University, Melbourne, VIC 3800, Australia; pkou0003@student.monash.edu
[2]   School of Business, University of New South Wales, Canberra, ACT 2612, Australia; e.chang@adfa.edu.au
[3]   Fleetwood Corporation Limited, Perth, WA 6004, Australia; stevew@glyde.net.au
*   Correspondence: m.yuzhang@unsw.edu.au

Abstract: Reducing energy waste is one of the primary concerns facing Remote Industrial Plants (RIP) and, in particular, the accommodations and operational plants located in remote areas. With the COVID-19 pandemic continuing to attack the health of workforce, managing the balance between energy efficiency and Occupation Health and Safety (OHS) in the workplace becomes another great challenge for the RIP. Maintaining this balance is difficult mainly because a full awareness of the OHS will generally consume more energy while reducing the energy cost may lead to a less effective OHS, and the existing literature has not seen a system that is designed for the RIPs to conserve energy usage and improve workforce OHS simultaneously. To bridge this gap, in this paper, we propose an AI Empowered Cyber Physical Ecosystem (AECPE) solution for the RIPs, which integrates Cyber-Physical Systems (CPS), artificial intelligence, and mobile networks. The preliminary results of lab experiments and field tests proved that the AECPE was able to help industries reduce the corporate annual energy cost that is worth millions of dollars, optimise the environmental conditions, and improve OHS for all workers and stakeholders. The implementation of the AECPE can result in efficient energy usage, reduced wastage and emissions, environment-friendly operations, and improved social reputation of the industries.

Keywords: cyber-physical system; ecosystem; remote industries; OHS; energy efficiency; smart meter; artificial intelligence; COVID-19

Citation: Koutroumpinas, P.; Zhang, Y.; Wallis, S.; Chang, E. An Artificial Intelligence Empowered Cyber Physical Ecosystem for Energy Efficiency and Occupation Health and Safety. *Energies* **2021**, *14*, 4214. https://doi.org/10.3390/en14144214

Received: 17 April 2021
Accepted: 8 July 2021
Published: 12 July 2021

Publisher's Note: MDPI stays neutral with regard to jurisdictional claims in published maps and institutional affiliations.

1. Introduction

Maintaining a delicate balance between energy efficiency and Occupation Health and Safety (OHS) [1] has been a challenge facing large enterprises and particularly the Remote Industrial Plants (RIP), such as mining campsites and other resources industries and remote industrial facilities in regional and countryside areas. With extreme weather conditions around the world, energy powered air conditioning systems need to be properly managed in order to prevent uninhabitable situations from happening.

As examples: when the temperature outside of a mining campsite is around 40 °C, it could exceed 50 °C indoors; one person working after hours turning on the entire site's or building's power and lights instead of the one specific spot that is needed; and malfunctioning air-conditioning units in an operational plant could lead all staff to experience stress due to high or low temperatures. The extreme weather conditions require a constant supply of energy with balanced heating or cooling to ensure that OHS standards are met for the sake of all employees and visitors as well as stakeholders.

In addition, the challenge extends to the issues of appropriate lighting, ventilation, air-quality, and humidity control. Regardless of whether the majority of RIPs are reliant on fossil fuels to generate power and are off-site, off-grid with no electrical networks, or in the city where the lights are on in continuous and constant mode. With massive energy consumption, the difficulty lies in how to properly manage the energy cost whilst ensuring OHS conditions are optimal for the workforce.

In order to address the above issues and risks in the RIPs, this paper presents an AI Empowered Cyber Physical Ecosystem (AECPE), which is a cost-effective, smart digital ecosystem solution that integrates Cyber-Physical Systems (CPS), mobile networks, and artificial intelligence. The AECPE is designed to intelligently provide personalised services, including a live-able temperature, refrigerating perishables, water temperature control, electrical appliances, lighting, ventilation, and room humidity, by detecting the occupancy of rooms or facilities and monitoring the habitable environments.

This differs from any existing CPS or Internet of Things (IoT) since its focus is on making the computing elements of the system coordinate more closely with the sensors and the actuators to achieve a higher level of system intelligence and efficiency in monitoring and controlling the cyber and physical environments. In addition, the mobile networks and AI enable the system and devices to exchange information, analyse data and produce real-time energy usage and environmental awareness. The AECPE leverages communications among sensors, smart meters, and AI modules through mobile networks to create a cyber physical ecosystem that not only reduces electricity wastage and carbon emissions but also fulfils OHS compliance at the same time.

The contribution of the proposed AECPE lies more towards its practical applications than theoretical level, which is summarised from three aspects. Firstly, more OHS profile entries are considered in the AECPE since the system is designed for remote area industrial plants and accommodations, including indoor/outdoor and historical/forecast temperature, lighting, air quality, ventilation, humidity, personal use preferences, and movement from area to area. Secondly, the actuators are managed by machine-learning based algorithms that require minimal human intervention, and that take into account valuable real data collected from test sites, such as user preferences, room occupancy and personal movement which can be used for not only maintaining the balance between energy efficiency and OHS but also contact tracing to offer proactive responses during the COVID-19 pandemic. Finally, the ACEPE has been tested on real data in a real-life environment with highly demanding accommodation and a physically stressful outdoor environment, and the results have shown success in reducing the electricity consumption while maintaining a user-friendly living environment.

The remainder of this paper is structured as follows. Section 2 presents the practical issues facing the current industries and technologies. Section 3 introduces the framework of the proposed AECPE. Section 4 describes the design of the AECPE in detail, followed by the prototype setup and field testing of the system in Section 5. Lastly, Section 6 presents the conclusions and directions for future study.

2. Related Work

2.1. Cyber-Physical Systems

A Cyber Physical System (CPS) is a system that provides new ways for humans to interact with the cyber and the physical world [2]. A CPS generally contains two main parts. The first is real-time connectivity to the physical world allowing continual information feedback, and the second is intelligent data management and analytics. Instead of using a two-part structure, a five-part structure of CPS was proposed [3]. They are (i) smart connection, (ii) data-to-information conversion, (iii) a cyber level, (iv) cognition level, and finally (v) configuration level. Each of these five sections will have subsections.

A smart connection should involve sensors and metering wireless networks and plug and play connectivity. This section does not have to be online if not necessary. A Zigbee protocol can be used for this section. Data analytics and desegregation will take part in the data-to-information section. There are two choices of how to go about this. The analytics can occur on-site where the data was collected or can be send to the cloud for the analytics to occur there. On-site analytics require for there to be a server framework created on said site. On the other hand, for the data to be send to the cloud could put a strain on devices if they are running on a battery or could put excess strain on the bandwidth available.

In a case study, it was seen that the power draw of the CPS system could be reduced by up to 70% if the collected data were first compressed before being sent to the cloud [4]. This promotes the idea that some amount of on-site pre-processing will have to occur. The cyber level will allow for further analysis to occur and models to be created to be able to identify similarities or variations within the incoming data. Next, the cognition level will allow the incoming data to be visualized for the user. Monitoring will play an important role in the future of CPS as it will enable real-time alerts and corrective actions depending on what is needed by the system [5].

The final level is the configuration level, which includes a large amount of machine learning algorithms so that the system can be self-adjustable, self-optimizing, and self-configuring. If the system can achieve these requirements, it will be able quickly adjust as the system grows and become more complex throughout its operation. These five steps can act as a good road map in the future implementation of new Cyber Physical Systems [3] .

In terms of CPS applications, these have been used in many sectors, such as aviation, IT, transport, and medical fields. The medical field has seen multiple case studies. One such study proposed the Medical Cyber Physical System or MCPS [6]. The main problems that a system like this would face are reliability, security, and safety requirements, which require further human interactions to obtain the best use out of the system. The paper also draws attention to the impact that cyber attacks can have as well as the need for a combined design methodology to resolve the design challenges of long-term learning and self-adaption of MCPS.

A further case study into the applications of CPS in the medical field was applied in a house environment [7] to improve the quality of life for elderly patients or other patients. What they proposed was the application of a camera and sensor network for real time surveillance of the patient to allow better and more efficient care. This would also allow real-time scheduling and management of resources—for example, the dispensing of medication. A point that this paper raised is aligned with another CPS research, which claimed that any lapses in the security and communication of the MCPS could put lives at risk [6]. The MCPS was also used to control an Analgesic Infusion pump [8].

Their application used a spatio-temporal partial differential equation and a PKA model to determine the concentration of a medication in a patient's blood stream. If the concentration drops too low, then the pump will be told to increase the output of medication and vice versa if the concentration is too high. This way, if there is a high enough confidence in the system, the human error can be removed, and the response time to the changes in concentration will be detected quicker by the software than by the human user, thus, improving the safety for the patient.

While CPS has been used as the framework to connect the cyber and physical worlds, many system control approaches have been proposed. An artificial bee colony optimisation algorithm was proposed to simultaneously maximise comfort whilst minimising energy usage [9]. Fuzzy controllers were used in the system where the input for the controllers was the difference between the input user parameters and the actual parameters, and the output is the required power to affect the conditions. This solution aims to address the Comfort index, which applies on the suitability of environmental conditions for physical activity. Applying fuzzy logic for system control has become more common as the need for energy saving is rising.

A fan coil unit was proposed as a way of transporting heat from the air into a coolant fluid [10]. In their experiment, the controlled variables were the room temperature and the room humidity. It was shown that fuzzy controls successfully managed the previous variables, whereas the Proportional Integral Derivative control failed to do so, which uses more energy. It can also be used for multiple system control, such as managing the room temperature control whilst taking into account the dew point [11]. This study showed that fuzzy logic was an easy way of solving this problem without getting too involved in the physical variables. Fuzzy logic has been shown to be an effective solution when dealing with a temperature control problem in certain situations.

However, fuzzy logic lacks the ability to make future predictions based on historical information, such as how much energy will be used in the future leading to future costs being known. Machine learning, on the other hand, excels at this given that a large amount of data is collected and ready for use [12]. In our project, three years of room temperature data was collected and provided for experiments, and the outdoor daily temperature for the region can also be accessed online [13], thus, making the machine learning a more effective solution than fuzzy logic for this situation.

2.2. Cyber Security

In any situation where a wireless connection is involved, cyber security is an issue that has to be taken into account. Different types of attack can occur. A denial-of-service (DoS) attack involves rendering a service, a computer, or other devices unavailable to the intended users by overwhelming a targeted machine or service with requests until normal requests cannot come through [14]. Sites, such as shopping centres, airports, universities, and other public locations, are vulnerable to these types of attacks due to their broadcasting nature [15].

These attacks can also slow down systems and make them use stale data [16]. In this case, it would drive up the energy usage as the optimisation algorithm relies on having the most current data. There is also a risk of the room not being at the correct temperature when the worker arrives, which would be an OHS risk. A CPS could also face replay attacks, which involve an intercepted message being delayed or intercepted to misdirect the receiver [17]. This attack could occur through the end sensors of the CPS, where the collected data could be collected and then duplicated [18].

A possible way to combat this attack is to use a random authentication signal in the control system [18] or encryption keys [17]. Another attack is a false data injection attack, which can happen in a power grid [19]. This attack also involves an attack on the sensors; however, in this case, it involves compromising the sensor reading in such a way that undetected errors are introduced into the calculations of state variables and values [20].

2.3. Occupational Health and Safety

Occupational Health and Safety (OHS) is the study of workplace influences on health and well-being, including identifying the environment and mechanism causes of imbalances related to human biological, chemical, and physical pollutants in the workplace and providing a customised and occupational health and safety compliant environment for workers, including remote operations. The Occupational Safety and Health Act and OHS risk management supported by the Occupational Safety and Health Regulations in each country are used to ensure that habitable conditions are met for the workers.

Occupational health is as important as physical health, and we developed mechanisms for better occupational health using data captured through CPS devices to infer and identify risk factors (i.e., stress, fatigue, and mental illness). There have been some studies, applications, and devices focusing on CPS (listed as follows); however, they all showed their limitations in considering the OHS.

- Smart metering from EKM Metering (US) and Meazon (EU) are great at monitoring energy and water usage to extremely high accuracy but they do not offer services to improve OHS [21].
- Paxton (UK) offers wireless door handles for access control but does not consider OHS [22].
- KBE, one of the world's largest manufacturers for energy saving buildings, focuses only on materials used for housing but not OHS [23].
- USA, China, and EU have worked on energy optimisation through the IoT coupled with renewable energy for remote and rural communities to improve lifestyle through energy supply, but do not focus on the corporate social responsibility for OHS of their workers or tenants [24,25].

- Smart Cities and smart homes are other areas close to this proposed project. An application of Smart Cities [26] to remote areas investigated improve lifestyle through energy efficient appliances but not OHS in the corporate environment.

3. AI Empowered Cyber Physical Ecosystem (AECPE) Solution Architecture

The AECPE solution architecture is proposed based on four main sections, namely OHS Profile data acquisition, ML-based optimiser, a smart metering system, and cyber security and COVID-19 response, to be able to fulfil the needs of energy efficiency, OHS requirements, and COVID-19 responses. The overall architecture is demonstrated in Figure 1 in which the sections are framed by dashed lines in different colours.

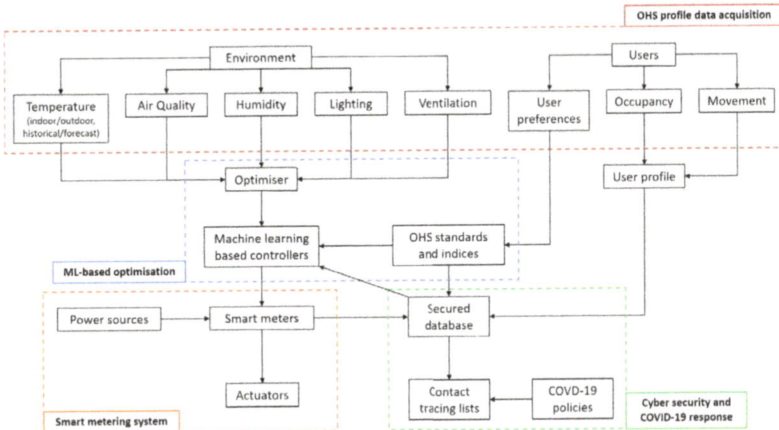

Figure 1. The AI Empowered Cyber Physical Ecosystem (AECPE) solution architecture.

Specifically, the environment data and user information that concern the OHS requirements are collected as an OHS profile input in the OHS profile data acquisition section. The environment profile is entered to an optimiser, and the user preferences together with the OHS standards and indices, are fed to the machine-learning-based controllers for a series of analysis, optimisation, and prediction in the ML-based optimiser. These results are input to the smart metering system in which the smart meters request the required electricity from the power sources and, meanwhile enable the actuators to switch the devices on and off (air conditions, lighting, ventilation, etc.).

All the data mentioned above is stored in a secured database, and this database can also talk to the ML-based controllers for continuous learning. On the other hand, the room occupancy and user movement are also monitored to create a user profile, which is also stored in the database for COVID-19 responses in terms of contact tracing according to COVID-19 policies. The details of these sections will be described later.

4. AECPE Design

4.1. OHS Profile Data Acquisition

The OHS profile is collected through monitoring the site environment and users in real time for correct decision making, for both human decisions and machine decisions. To this end, the system will need to be able to access any stored data or any processed data. Middle-ware will also need to be developed and applied over the smart gateway system to help with management of the network and to make any monitoring applications easier to run. Monitoring should also be able to allow for the use of regression algorithms combined with time series algorithms that can predict future energy usage using the past data. This will allow for the better distribution of resources leading to reductions in operations costs. Information to help with COVID-19 prevention can also be collected as part of the OHS Profile.

The OHS profile includes indoor and outdoor temperature, as well as temperature in the past and climate forecast for the near future, air quality, humidity, lighting, ventilation, personal user preferences, room occupancy, and user movement.

Having a room at a comfortable temperature is important, as heat stroke can occur easily in an environment that is under a near constant heat wave. A heat wave is defined as an area whose temperature over 32.2 °C for 3 days or more in a row, which is a criteria that is met by the chosen site for multiple months every year [27]. Having this heat wave period greatly increases the risk of heat stroke occurring. Heat stroke is when the body's core temperature goes above 40 °C [27]. The risk of this occurring is even higher when workers return after a 10-h shift of physical labour. Where the temperature of the room is kept exactly will be subject to the user when they are in the room. This is done through either the remote control of the air-conditioners for a one time change or by setting a new permanent temperature through the app.

Another aspect that needs to be accounted for is the dew point and humidity of the area, which changes greatly over time as there are large swings in the daily temperature and the relative humidity. The dew-point must be precisely controlled. This is due to the fact that it has been observed (on site) that, if the temperature in the room is left beneath the dew point, and moisture starts to build and mould can occur. If mould builds in a room and it is not cleaned fast enough, it can cause permanent structural damage.

It can cost up to 2000 dollars to have mould cleaned, and it can render a block of four apartments unlivable for up to 2 weeks. Therefore, whilst taking into account that the temperature has to be kept cool enough to be lived in, the room also has to be allowed to be heated above the dew point temperature long enough for moisture to evaporate to stop mould from forming. Controlling humidity will also help to ensure that the air quality in the rooms is maintained at a good level.

The capture kit containing the sensors and actuators that allow OHS to be monitored was installed in the accommodation to start collecting data. This captured data plays a crucial role in delivering the balance between energy efficiency and OHS outcomes. A secure communication between the sensors, the kit, and the cloud servers was established with security mechanisms to maintain data security and to ensure the users' privacy.

Sample temperature data from the whole site was collected by both indoor and outdoor thermometers, including individual room thermometers and inbuilt air-conditioning thermometers as well as checked against online meteorological data for the region. Both indoor and outdoor temperatures are required, as there is a tendency for the internal temperature to be higher than external temperatures. At higher temperatures, there is a strong correlation between the two temperatures [28] that is not there at lower temperatures.

As the site is located in a region that has average maximum temperatures reaching above 30 °C for six months of the year and mid to high 20 °C for the other half year [29], this linear correlation occurs from when the temperature is greater than 12.7 °C, which is well below even the average minimum temperature for the site. Collection and analysis of this data will lead to a better understanding of localised energy usage and OHS needs. The third stage will provide the roster and demographic data. This step also includes pre-process data that includes data cleansing, data transformation, and labelling process. After pre-processing, data are fed into the AI and machine learning algorithms for analysis.

The overall data collection and analysis are demonstrated in Figure 2. This flowchart shows how incoming data from energy usage and sensor data is collected and processed, to generate reports and generate new commands to be able to run a more efficient site.

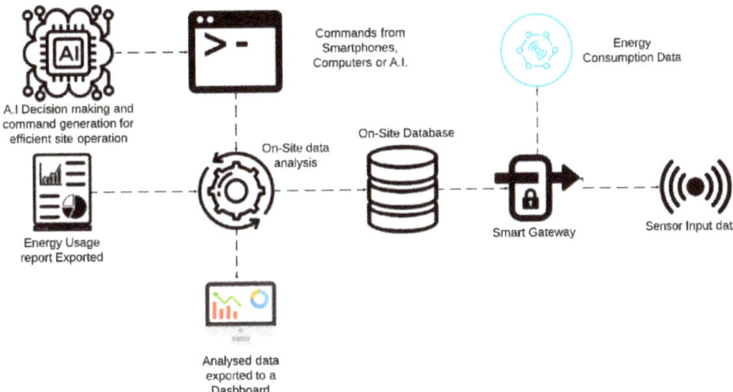

Figure 2. Data collection and analysis.

4.2. Smart Metering System

Integrating smart meters within a mesh network will allow for a flexible system that can be monitored and controlled in real-time to meet any needs that a site might require. Monitoring the energy consumption and system performance will allow for precise device scheduling to manage energy consumption while retaining health and well-being conditions at its most optimal. This will be done by analysing rosters for people coming into the plant or accommodation sites. From those rosters, the arrival and departure dates of the workers will be extracted along with the shifts that they will be working, through which the indoor temperature will be managed to be as efficient as possible.

This will be done by taking environmental data into account, such as the external temperature and humidity; as well as room data, such as the internal temperature and time until the return of the worker to the room. Energy consumption monitoring and scheduling techniques aim to reduce overall energy consumption without compromising the level of comfort. Two-way communication allows operation control over appliances, hence, enabling the operation of energy consumption scheduling. The techniques incorporate the aggregated demand to optimise electricity consumption across the plant and accommodation sites. The design of the AECPE physical layout is demonstrated in Figure 3.

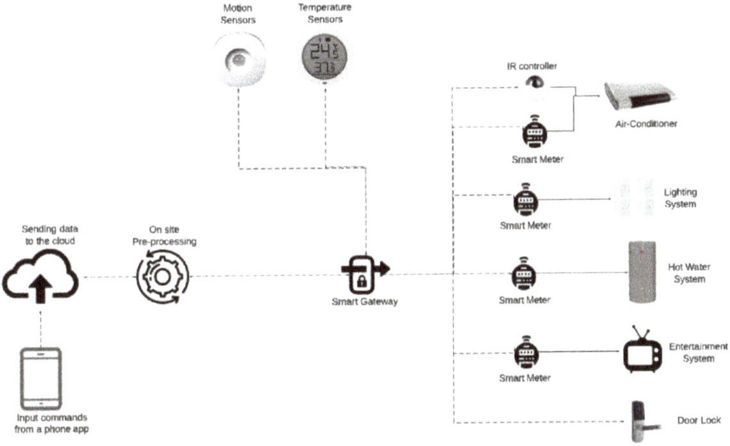

Figure 3. The on-site layout of devices and smart meters.

In this figure, the dotted lines represent wireless connections, whether they are Zigbee connections between the gateway and the meters and sensors or Ethernet connections between the gateway and pre-processing. The pre-processing that will occur is to combine the different data sets of energy usage and weather conditions so that the regression algorithm can act on them. The solid lines represent actual wiring connections where electricity is fed to each of the machines or systems. Zigbee is a good choice for the device control section of the system as it is a low energy use protocol, which means that a device can run on a battery for a long time. Zigbee can also connect to up to 4160 plus devices [30], which means that the mesh network can be stretched to cover any size site that may be required.

4.3. ML-Based Optimisation

Machine learning techniques were used to analyse the collected sensor data and OHS profiles to maintain the energy efficiency and workplace environment. Linear Regression models allow cause–effect relationships to be identified using past energy usage data combined with daily temperature information collected online. This will be done so that it is known at what temperature setting the air conditioning units need to be running for the most efficient energy usage based on the outdoor temperatures and required indoor temperature. This will result in energy saving while maintaining the health and well being for occupants.

Time-series forecasting algorithms will also be employed to enable energy consumption scheduling (the next milestone) and adaptive actions for health and well-being (mainly keeping the temperature to a comfortable level). Time-series forecasting will also allow an estimation of future energy consumption and future energy costs for informed decision making. The formation of mould is another risk that needs to be mitigated. Mould forms when the temperature of the room is kept below the dew point for a consistent period. If this happens, the kit will adjust to increase the temperature to stop mould formation—provided it does not go against any OHS requirements.

A regression algorithm predicts the output values based on the input features from the data fed into the system [31]. Specifically, a multivariate regression model will be trained to explain the relationship between the energy consumption and the OHS profiles of the room (e.g., the indoor temperature, outdoor temperature, and past and forecast temperature), while time series models will be used to predict the future temperature. For the regression model, the overall energy cost of a room for a whole day is set to be the target variable, and this will be learnt and optimised based on the OHS profiles, which are used as independent variables. The prediction of the output values will also allow for forecasting to occur. Future cost prediction will be made by correlating past costs with the predicted future energy usage to find future costs.

The general form of the linear regression model, given a data set $\{y_i, x_1, \ldots, x_i\}_{i=1}^{n}$ for n independent variables, can be written as:

$$y_i = \beta_0 + \beta_1 \cdot x_1 + \ldots + \beta_i \cdot x_i + \ldots + \beta_n \cdot x_n + \varepsilon \tag{1}$$

where $x_1, \ldots, x_n$ are the independent variables, $\beta_0, \ldots, \beta_n$ are the regression coefficients, y_i is the estimated outcome variable, and ε is the error capturing the other factors that influence the y other than those considered within $\{x_i\}_{i=1}^{n}$.

In this paper, the variables were constructed as in Figure 4, in which we consider the following independent variables as input, including the maximum/minimum outdoor/indoor temperature of the past five days ($OTmax_i, OTmin_i, ITmax_i, ITmin_i, i \in \{i = -5, -4, \ldots, -1\}$), the outdoor temperature in the past five hours ($OT_j, j \in \{i = -5, -4, \ldots, -1\}$), the preferred room temperature of the user in the past five hours ($PT_j, j \in \{i = -5, -4, \ldots, -1\}$), and the current indoor temperature (IT_{crt}), outdoor temperature (OT_{crt}), and current preferred room temperature (PT_{crt}).

The time series models were used to predict the indoor temperature (IT_{pdt}), outdoor temperature (OT_{pdt}), and preferred room temperature (PT_{pdt}) based on the corresponding variables in the past days and hours. Lastly, all the collected and predicted data were fed into a regression model as independent variables, and the dependent variable was the overall energy cost of a room for a day, including the time when the room was used and not used.

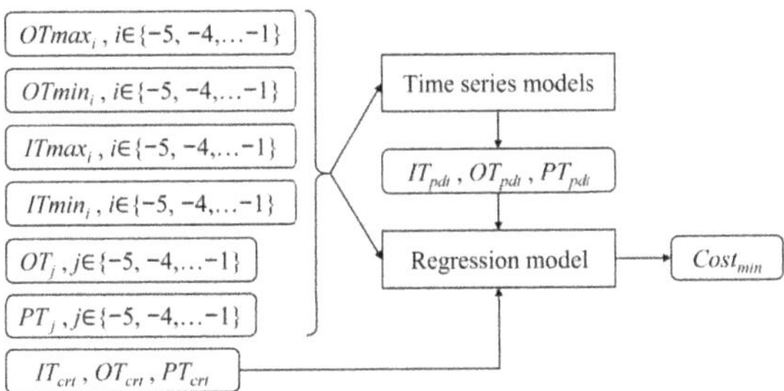

Figure 4. Machine learning models. The inputs and outputs of the models are in the rounded rectangles.

The cost prediction needs to give multiple estimates to account for a best to worst case scenario depending on how much variation is in the previous values that were used in the cost calculation. The aim is to set the IT_{crt} at a dynamic temperature to achieve the minimised *Cost*. In order to search for the optimised thermostat setback temperature for each room, we used the historical outdoor temperature and initiated the IT_{crt} and PT_{crt} using random numbers within a reasonable range to start up the training progress.

As for the IT_{crt}, it was set within a range from 17 to 33 °C, and the PT_{crt} was set within 22 to 28 °C, because the historical data showed that most users set the air conditioner at 22 to 28 °C after coming back to the room, and 5 extra degrees to both ends were extended for the air conditioner to handle extreme climates in the winter and summer. N, the number of IT_{crt}, and M, the number of PT_{crt}, were randomly generated for initial training, and the room occupation hours were also random numbers reflecting that the users could come back to the room at any time of day. The room needed to reach the PT_{crt} within 20 min after the PT_{crt} was set. Algorithm 1 summarises the details to search for the optimised energy cost.

Algorithm 1 Energy optimisation algorithm

Input: $OTmax_i$, $OTmin_i$, $ITmax_i$, $ITmin_i$, OT_j, PT_j, IT_{crt}, OT_{crt}, PT_{crt},
     $i \in \{i = -5, -4, ..., -1\}$, $j \in \{i = -5, -4, ..., -1\}$
Output: IT_{crt}, $Cost_{min}$
Parameter: $\beta_0, ..., \beta_n$
1   Initialise: $IT_{crt,p} \leftarrow a$, $PT_{crt,q} \leftarrow b$, $a \in \{17, ..., 33\}$, $b \in \{22, ..., 28\}$
2   Train the regression model: $Cost = \beta_0 + \beta_1 \cdot OTmax_i + ... + \beta_n \cdot PT_{crt} + \varepsilon$
3   **for** $p = 1 : N$ **do**
4   **for** $q = 1 : M$ **do**
5   **if** $Cost_m - Cost_{m-1} \leq 0$ **then**
6   **return** $IT_{crt,p}$
7            $Cost_{min} \leftarrow Cost_m$
8   **return** $Cost_{min}$

4.4. Cyber Security and COVID-19 Response

In the AECPE, security is handled by a set of firewalls on top of the systems being accessible only via a VPN. For the moment, whilst the system is still in development and only deployed very locally within one accommodation site, this security has proven to be sufficient; however, it will need to be improved with a proper cyber threat response system when it is ready to deploy to more locations. The security will mostly focus on the protection of the end sensors/actuators as was outlined in the replay attack section.

As for the COVID-19 response, the movement of the people around the site can be tracked using their ID cards when they tap them to gain access to their rooms and to other common facilities. In this way, if a case of COVID-19 is detected, this person's close contacts and the areas he or she has been in during any requested time period can be quickly queried via the system. The system can also be customised to include temperature checks to scan on the arrival of the workers and other random checks throughout their time on site.

A body temperature threshold (37.8 °C) will be set up to trigger the system to start recording a person's temperature, and an alarm will be sent to this person and campsite manager if a body temperature above 38.28 °C [32] is detected; meanwhile, the user profile (movement and occupancy) of this person will be flagged in the system and ready to conduct contact tracing for a fast COVID-19 response.

5. Prototype Validation and Field Testing

5.1. AECPE Laboratory Prototype Setup

A prototype system was set up and tested in our lab environment in Australia. Four different devices were connected onto the system at the one time as demonstrated in Figure 5. The system could then measure how much electricity was being used and also provided remote control of the devices. On a small scale, it was shown that the system operated as was expected. Further research will show if this control functionality can be used within the system when automated commands are given instead of user commands. It will also be interesting to determine to what extend feedback loops can be used and when a user needs to step in to give new commands.

Figure 5. The lab testing setup where four plugs and a gateway (on the right) are connected to a laptop, a screen, and two desk lamps. Their energy usage data was collected and remote control was established over those devices.

5.2. AECPE Field Testing

We further conducted validation of the AECPE prototype in field testing in a large mining campsite provider and manager enterprise in West Australia. Specifically, the features of continuously optimising energy usage and OHS awareness were tested in one of the campsites provided and managed by the enterprise. Since this campsite is composed of a considerably large size of more than 1000 units meaning more than 3000 plus devices were connected, we also investigated the communication amongst sensors and smart meters, as well as how much the mesh network could be stretched.

In the field testing, the AECPE mainly monitored workers in and out of mining camp/room for the optimal use of electricity, and this was trialled in over 600 rooms in the mining campsite. The preliminary results demonstrated that the energy costs could be reduced by 50% within each camp as the system could manage the energy usage at each individual occupancy zone rather than control the energy usage in parts of the complex or part of the site/camp/building/floor/area to reduce energy waste.

In the current situation, all the appliances operated all the time regardless of the occupancy, which led to a significant amount of energy waste. Each kilowatt per hour of energy costs $0.30 and produces approximate 1 kilo of carbon dioxide every day. If a typical mining campsite has 1000 rooms/units, with a savings of $1 per room/per day, the AECPE could save at least $365,000 p.a./per camp. These savings happen through only running the air-conditioning as much as is needed. Testing has shown that it tends to be more efficient to maintain the room at a temperature ranging from 24 °C to 30 °C.

This is due to the inside room temperature getting so high (possibly up to 50 °C) that it takes more energy to cool the room down to 19 °C again from that temperature instead of the amount of energy that required to maintain 28 °C and then cool down to 19 °C. For early testing, this was done by creating a few preset temperature states and preset scheduling. The AI algorithm has to make real time decision with the incoming temperature and roster data to ensure that the room is at the required temperature when the worker returns and also for what temperature to hold the room at while the worker is away.

Australia currently runs over 350 such mine sites [33], which all have accommodation camps associated with them. As of 2020, there were 261,900 miners working in the mining industry in Australia [34]. Applying the AECPE will result in millions of dollars saved, reduced energy consumption, and minimised carbon emissions. These figures would result in a reduction of approximately 1200 tonnes of carbon dioxide emissions p.a./per camp. This calculation is only considering the habitable facility and the mining camp and not the corporate buildings, floors, or areas.

Figures 6 and 7 demonstrate a session of our field testing where the power consumption of four different air-conditioning units in four rooms was tested and recorded over April and October reflecting two different seasons in the remote areas of Australia. The solid line refers to the unit with the proposed AECPE installed for its temperature and energy usage control, and the dashed lines denote those without instructional operation or relying on a rudimentary energy saving algorithm. Overall, AECPE had a comparatively lower energy usage than the other air-conditioning units. Specifically, it used 40% of AC1's power consumption, 47% of AC2's power consumption, and 46% of AC3's power consumption on average throughout the month.

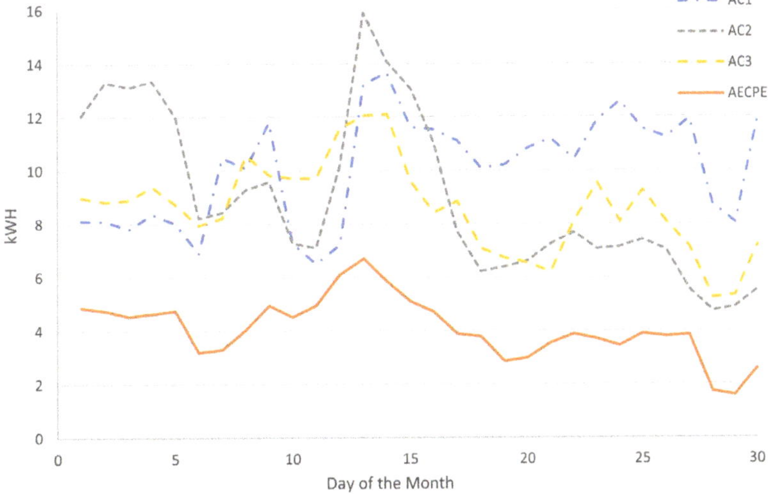

Figure 6. April results from the field test.

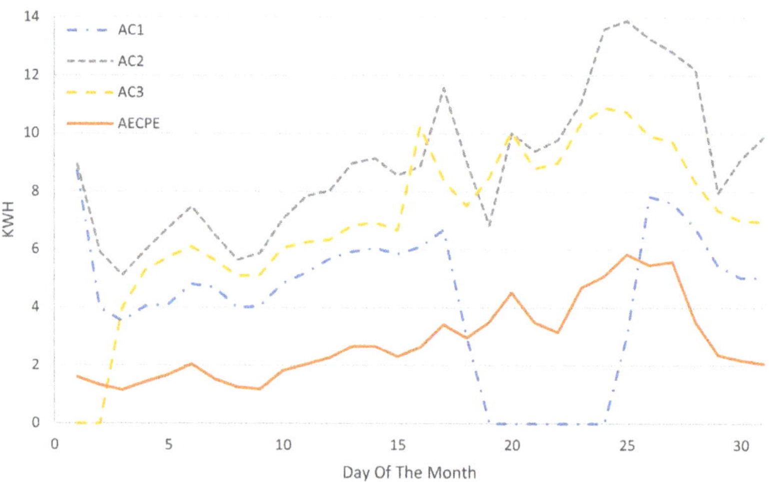

Figure 7. October results from the field test.

6. Conclusions

We developed an AI-Empowered Cyber-Physical Ecosystem (AECPE) that is capable of monitoring energy consumption in real time and maintaining OHS parameters. This system aims to assist in achieving energy efficiency and better OHS outcomes for workers and facilities in remote regions. The combined energy and OHS technology will help remote operations in not only energy waste reduction but also a better understanding of energy usage and OHS needs. A CPS- and AI-based platform was designed to provide integrated analytics enabling the efficient management of facilities, including energy consumption, improved OHS, emission reductions, efficient resource utilisation, and efficient facility maintenance, along with timely decision support between differing autonomous CPS devices, including smart metering systems.

The AECPE is based on the integration of smart metering and machine-learning-based optimisation with OHS. Bringing these elements together is important as it will allow for the system to run in its most efficient state. This combination will allow this system to provide a new way of controlling and overseeing the OHS of a site over a mobile phone. Using a new generation of smart meters, it will allow measurements to be taken at a faster speed than any other smart meters (as fast as one every 20 ms) with an accuracy of 99%.

Measurements taken this frequently will require a greater amount of storage if the internet connection is to ever drop out at a remote site. The gateway will provide this storage, as it can store the data from multiple days until it becomes reconnected to the internet. This would allow for no loss of data to occur to ensure accurate analysis and future predictions. The system is also capable of responding to the COVID-19 pandemic in terms of user monitoring and contact tracing, which is aligned with the government policies regarding the OHS during this pandemic period.

For future work, we will work on developing machine learning and predictive analytical algorithms for pattern recognition and integrated analytics for the two interlinking areas, namely energy consumption and OHS, to help optimise energy usage and eliminate waste while maintaining health and well-being standards. In addition, we will also provide energy and OHS training programs for senior managers in the remote mining organisations and assist them with low-cost smart mining camp solutions based on our collected data and analysis results. The cyber security system will also need to be updated as the work goes along to keep up with potential threats.

Author Contributions: Conceptualization: P.K., S.W. and E.C.; Methodology: P.K., Y.Z. and E.C.; Software: P.K.; Validation: P.K. and S.W.; Investigation: P.K. and Y.Z.; Resources: P.K., S.W. and E.C.; Writing—original draft preparation: P.K.; Writing—review and editing: P.K., Y.Z. and E.C.; Supervision: Y.Z. and E.C. All authors have read and agreed to the published version of the manuscript.

Funding: This research received no external funding.

Conflicts of Interest: The authors declare no conflict of interest.

References

1. Stellman, J.M. *Encyclopaedia of Occupational Health and Safety*; International Labour Organization: Geneva, Switzerland, 1998; Volume 1.
2. Chang, E.; Gottwalt, F.; Zhang, Y. Cyber situational awareness for CPS, 5G and IoT. In *Frontiers in Electronic Technologies*; Springer: Singapore, 2017; pp. 147–161.
3. Lee, J.; Bagheri, B.; Kao, H.A. A cyber-physical systems architecture for industry 4.0-based manufacturing systems. *Manuf. Lett.* **2015**, *3*, 18–23. [CrossRef]
4. Horcas, J.M.; Pinto, M.; Fuentes, L. Context-aware energy-efficient applications for cyber-physical systems. *Ad Hoc Netw.* **2019**, *82*, 15–30. [CrossRef]
5. Bartocci, E.; Deshmukh, J.; Donzé, A.; Fainekos, G.; Maler, O.; Ničković, D.; Sankaranarayanan, S. Specification-based monitoring of cyber-physical systems: A survey on theory, tools and applications. In *Lectures on Runtime Verification*; Springer: Cham, Switzerland, 2018; pp. 135–175.
6. Dey, N.; Ashour, A.S.; Shi, F.; Fong, S.J.; Tavares, J.M.R. Medical cyber-physical systems: A survey. *J. Med. Syst.* **2018**, *42*, 74. [CrossRef] [PubMed]
7. Wang, J.; Abid, H.; Lee, S.; Shu, L.; Xia, F. A secured health care application architecture for cyber-physical systems. *arXiv* **2011**, arXiv:1201.0213.
8. Banerjee, A.; Gupta, S.K.; Fainekos, G.; Varsamopoulos, G. Towards modeling and analysis of cyber-physical medical systems. In Proceedings of the 4th International Symposium on Applied Sciences in Biomedical and Communication Technologies, Barcelona, Spain, 26–29 October 2011; pp. 1–5.
9. Wahid, F.; Kim, D.H. An Efficient Approach for Energy Consumption Optimization and Management in Residential Building Using Artificial Bee Colony and Fuzzy Logic. *Math. Probl. Eng.* **2016**, *2016*, 9104735. [CrossRef]
10. Attia, A.H.; Rezeka, S.F.; Saleh, A.M. Fuzzy logic control of air-conditioning system in residential buildings. *Alex. Eng. J.* **2015**, *54*, 395–403. [CrossRef]
11. Patanaik, A. *Fuzzy Logic Control of Air Conditioners*; Indian Institute of Technology: Kharagpur, India, 2008.
12. Kale, S.S.; Patil, P.S. Data mining technology with fuzzy logic, neural networks and machine learning for agriculture. In *Data Management, Analytics and Innovation*; Springer: Singapore, 2019; pp. 79–87.
13. Past Weather in Karratha, Western Australia, Australia. Available online: https://www.timeanddate.com/weather/australia/karratha/historic (accessed on 4 June 2021).
14. Hasbullah, H.; Soomro, I.A. Denial of service (DOS) attack and its possible solutions in VANET. *Int. J. Electron. Commun. Eng.* **2010**, *4*, 813–817.
15. Bicakci, K.; Tavli, B. Denial-of-Service attacks and countermeasures in IEEE 802.11 wireless networks. *Comput. Stand. Interfaces* **2009**, *31*, 931–941. [CrossRef]
16. Krotofil, M.; Cárdenas, A.A.; Manning, B.; Larsen, J. CPS: Driving Cyber-Physical Systems to Unsafe Operating Conditions by Timing DoS Attacks on Sensor Signals. In *Proceedings of the 30th Annual Computer Security Applications Conference (ACSAC '14)*; Association for Computing Machinery: New York, NY, USA, 2014; pp. 146–155. [CrossRef]
17. Nagarsheth, P.; Khoury, E.; Patil, K.; Garland, M. Replay Attack Detection Using DNN for Channel Discrimination. In Proceedings of the Interspeech 2017—18th Annual Conference of the International Speech Communication Association, Stockholm, Sweden, 18–24 August 2017; pp. 97–101.
18. Hosseinzadeh, M.; Sinopoli, B.; Garone, E. Feasibility and Detection of Replay Attack in Networked Constrained Cyber-Physical Systems. In Proceedings of the 2019 57th Annual Allerton Conference on Communication, Control, and Computing (Allerton), Monticello, IL, USA, 24–27 September 2019; pp. 712–717. [CrossRef]
19. Liu, Y.; Ning, P.; Reiter, M.K. False Data Injection Attacks against State Estimation in Electric Power Grids. *ACM Trans. Inf. Syst. Secur.* **2011**, *14*. [CrossRef]
20. Ahmed, M.; Pathan, A.S.K. False data injection attack (FDIA): An overview and new metrics for fair evaluation of its countermeasure. *Complex Adapt. Syst. Model.* **2020**, *8*. [CrossRef]
21. Mpelogianni, V.; Groumpos, P.; Tsipianitis, D.; Papagiannaki, A.; Gionas, J. Proactive Building Energy Management based on Fuzzy Logic and Expert Intelligence. *Inform. Intell. Syst. Appl.* **2020**, *1*, 56–58. [CrossRef]
22. Paxton. Access Control. Available online: https://www.paxton-access.com/solutions/access-control/ (accessed on 8 April 2021).
23. KBE Building Corporation. Available online: https://kbebuilding.com/portfolio/ (accessed on 8 April 2021).
24. Zhang, Y.; Guo, Z.; Lv, J.; Liu, Y. A Framework for Smart Production-Logistics Systems Based on CPS and Industrial IoT. *IEEE Trans. Ind. Inform.* **2018**, *14*, 4019–4032. [CrossRef]

25. Törngren, M.; Asplund, F.; Bensalem, S.; McDermid, J.; Passerone, R.; Pfeifer, H.; Sangiovanni-Vincentelli, A.; Schätz, B. Characterization, analysis, and recommendations for exploiting the opportunities of cyber-physical systems. In *Cyber-Physical Systems*; Elsevier: Amsterdam, The Netherlands, 2017; pp. 3–14.
26. Hernández-Muñoz, J.M.; Vercher, J.B.; Muñoz, L.; Galache, J.A.; Presser, M.; Gómez, L.A.H.; Pettersson, J. Smart cities at the forefront of the future internet. In *Future Internet Assembly*; Springer: Berlin/Heidelberg, Germany, 2011; pp. 447–462.
27. Bouchama, A.; Knochel, J.P. Heat Stroke. *N. Engl. J. Med.* **2002**, *346*, 1978–1988. [CrossRef] [PubMed]
28. Nguyen, J.L.; Schwartz, J.; Dockery, D.W. The relationship between indoor and outdoor temperature, apparent temperature, relative humidity, and absolute humidity. *Indoor Air* **2014**, *24*, 103–112. [CrossRef] [PubMed]
29. Karratha Climate. Available online: https://en.climate-data.org/oceania/australia/western-australia/karratha-11105/ (accessed on 29 May 2021).
30. Zigbee Pro Best Practises. Available online: https://www.control4.com/docs/product/zigbee/best-practices/english/revision/A/zigbee-best-practices-rev-a.pdf (accessed on 29 May 2021).
31. Binder, J.J. On the use of the multivariate regression model in event studies. *J. Account. Res.* **1985**, *23*, 370–383. [CrossRef]
32. What Temperature Is Considered a Fever? Available online: https://www.singlecare.com/blog/fever-temperature/ (accessed on 14 June 2021).
33. Australian Mineral Facts. Available online: https://www.ga.gov.au/education/classroom-resources/minerals-energy/australian-mineral-facts#:~:text=Source%20Geoscience%20Australia.,from%20over%20350%20operating%20mines (accessed on 28 May 2021).
34. In Numbers: How Mining Came to be Australia's Most Profitable Sector. Available online: https://www.mining-technology.com/features/in-numbers-how-mining-came-to-be-australias-most-profitable-sector/ (accessed on 28 May 2021).

Article

A Generic Pipeline for Machine Learning Users in Energy and Buildings Domain

Mahmoud Abdelkader Bashery Abbass [1,*] and Mohamed Hamdy [2]

[1] Department of Mechanical Power Engineering, Helwan University, Cairo 11772, Egypt
[2] Department of Civil and Environmental Engineering, Norwegian University of Science and Technology, 7491 Trondheim, Norway; mohamed.hamdy@ntnu.no
* Correspondence: Mahmoud.Gohar1992@m-eng.helwan.edu.eg

Abstract: One of the biggest problems in applying machine learning (ML) in the energy and buildings field is the lack of experience of ML users in implementing each ML algorithm in real-life applications the right way, because each algorithm has prerequisites to be used and specific problems or applications to be implemented. Hence, this paper introduces a generic pipeline to the ML users in the specified field to guide them to select the best-fitting algorithm based on their particular applications and to help them to implement the selected algorithm correctly to achieve the best performance. The introduced pipeline is built on (1) reviewing the most popular trails to put ML pipelines for the energy and building, with a declaration for each trial drawbacks to avoid it in the proposed pipeline; (2) reviewing the most popular ML algorithms in the energy and buildings field and linking them with possible applications in the energy and buildings field in one layout; (3) a full description of the proposed pipeline by explaining the way of implementing it and its environmental impacts in improving energy management systems for different countries; and (4) implementing the pipeline on real data (CBECS) to prove its applicability.

Keywords: machine learning; benchmarking; prediction; pipeline; features; training; validation; tuning; evaluation and model verification

Citation: Abbass, M.A.B.; Hamdy, M. A Generic Pipeline for Machine Learning Users in Energy and Buildings Domain. *Energies* **2021**, *14*, 5410. https://doi.org/10.3390/en14175410

Academic Editors: Francesco Nocera and Ana-Belén Gil-González

Received: 27 July 2021
Accepted: 29 August 2021
Published: 31 August 2021

1. Introduction

Building energy benchmarking and prediction is a complex (i.e., multi-variant and nonlinear) problem. Building energy demands depend on many features such as climate conditions, characteristics of a building, and the type of equipment in the building. The demands include electrical and thermal (heating and cooling) loads. ML algorithms can solve this type of problem as they automatically derive hidden patterns in the collected data. The patterns are then used to create the ML model, which generalizes real-life problems to provide more well-informed and adaptive results.

There are a lot of ML algorithms that are used in the energy and buildings field, but this paper explains the most popular algorithms that have dynamic behavior and are widely used in the field. Dynamic behavior means the ability of an algorithm to solve different problems in different applications, and the ability of algorithm integration with other algorithms to improve overall performance. The paper focuses on four ML algorithms: (1) artificial neural networks (ANNs), (2) support vector machine (SVM), (3) Gaussian process regression (GPR) or Gaussian mixture models (GMM), and (4) clustering (such as k-means and k-shape clustering algorithms).

To identify the essential steps required for implementing the ML concepts in the energy and buildings field, previous trials must be reviewed. In 2014, Zhao mentioned a pipeline for prediction energy values by split data into two data sets: (1) a training set that adjusts the weights of the ML model and (2) a test set to evaluate the trained ML model, without any data preprocessing. This technique is not enough to overcome drawbacks of (1) data quality such as missing values or outliers or noisy, and (2) overfitting training data because of adjusting ML model on the same data set, as well as the loss of some data in the test set not seen by the ML model [1]. In 2019, Tabrizchi, Javidi, and Amirzadeh Kim presented a prediction pipeline depending on the same two data sets but applying a cross-validation technique on the training data set to overcome the problem of overfitting. They proposed a pipeline depending on the feedback or results from the model evaluation process to make an optimization process for model parameters and a feature selection process, which help to reduce problem dimensionality without reducing ML model performance [2]. In 2019, Cai et al. declared the process of feature selection through a pipeline for the classification process in a layer called feature engineering which has also feature an extraction process, and dealing with missing data and outliers' values is also explained as a preprocessing layer [3]. The importance of the feature selection process and the case that is used in to be effective, is declared in the pipeline proposed in 2020 by Seyedzadeh et al., in addition to an explanation of the feature extraction process, which is very important when the algorithm cannot perform automatic feature extraction during training [4].

On other hand, Somu, Raman, and Ramamritham, in 2021, mentioned adding a third data set called the validation data set that is used to overcome the overfitting problem of ML models, but the problem of losing some data points while making the test data set remains. In addition to adding a preprocess layer containing processes of increasing data quality such as clean data from noise, missing values imputation, outlier detection, and data normalization, the authors also mentioned different evaluation methods for prediction problems such as mean square error (MSE), root mean square error (RMSE), mean absolute error (MAE), or mean absolute percentage error (MAPE). However, they added a benchmarking process as a feature extraction preprocess that integrates the clustering layer with the prediction pipeline to decrease the complexity of the prediction process [5]. The preprocess layer and evaluation layer were proposed before by Fayaz and Kim in 2018, but not in depth [6].

One of the trials of a general pipeline of ML algorithms was carried out by Liu et al. in 2019. The paper proposes different structures of pipelines, each one depending on the ML algorithm used inside. The layers of the proposed pipelines are data collection and preprocessing, training and evaluating models, and determining the best model parameters and structure. The authors mentioned a very important layer that must be found in the pipeline, especially in the implementation of a real-life problem. This layer is called the verification layer, which is very important to measure the trained model's robustness during operation. The weakness point of the proposed pipelines is that the authors cannot make a general pipeline cover all the requirements of different algorithms in different cases [7]. The trial of generating a general pipeline that suitable with different ML algorithms was performed by El-Gohary et al. in 2018. The pipeline deals with four ML algorithms: Naive Bayes, SVM, Decision Trees, and Random Forest. The pipeline depends on layers of data preprocessing, feature extraction, principal component analysis, and an evaluation layer. The main drawback of this pipeline is that it depends on only one path for any real-life problem, which cannot be generalized in all cases, and the authors do not mention requirements or criteria of selecting each ML algorithm. In addition, putting the classification process as a preprocessing layer to decrease prediction complexity means that the classification is an essential step [8]. In 2018, Saleh Seyedzadeh, Farzad Pour Rahimian, Ivan Glesk, and Marc Roper propose three pipelines: (1) a prediction pipeline depending on splitting data to train and test data which make a feature extraction process depending on the validation of model on test data, (2) a classification pipeline which depends on feature selection as a preprocessor process before classification, and (3) a pipeline to select the most appropriate ML algorithm (ANN, SVM, GPR or GMM, k-mean, and k-shape) depending on the data set and requirements of each algorithm. The drawbacks of pipelines are (1) neglecting the preprocessing process that must be performed to increase data quality, (2) depending on the simple technique of split data to train and test data, which cannot produce a general model solution, (3) the third pipeline does not cover all ML algorithms' requirements or data set cases, and (4) the authors cannot integrate the three pipelines into one general pipeline [9].

After reviewing the previous trials to create a generic ML pipeline, the resultant pipeline consists of main three general steps: (1) the preprocessing steps, (2) the ML algorithm selection, and (3) the ML model creation and implementation scientifically in real life. The most complicated part of creating this pipeline is the interaction between these three general steps (the main interactions come during ML selection procedures), so the proposed pipeline overcomes all obstacles for ML users. The ML algorithm selection step represents the main source of interaction between the three main steps and is usually difficult (i.e., selecting the most appropriate ML algorithm for a specific application and implementing it in real life) because it depends on many factors related to applications (e.g., energy assessment and forecasting; prediction for buildings loads such as cooling or heating or electricity; classification for buildings depending on energy consumption; modeling solar radiation; modeling and forecasting loads for air conditioning systems; simulating and control for energy consumption systems; fault detection and diagnosis; and energy-saving, verification and retrofit studies) and factors related to data (e.g., data size, features size, data type (residential or non-residential data and time serious or not); degree of uncertainty in data; and degree of complexity in data). After solving all interactions between the three main steps, the final pipeline structure covers several factors including problem formulation, data collection and integration, data augmentation, feature engineering, data preprocessing and visualization, different machine learning approaches with requirements, model training, model validation and tuning, model evaluation, and model verification. The generic ML pipeline will enhance the performance and organization of the reviewed ML algorithms, because while working on ML problems, many steps are heavily repeated, and thus, putting these steps into one generic pipeline will ensure that the right algorithms are deployed seamlessly, reducing the complexity of transferring ML models to real life quickly and managing ML models easier.

The proposed paper consists of two main sections besides the introduction section. Section 2 explains each step of the proposed pipeline, with some previous cases demonstrated (i.e., the most popular ML algorithms and their applications used in the building energy field and how each one is used to have most benefits). Section 3 implements the pipeline on CBECS data as an example to help ML users in using it.

2. The Essential Steps and Potential Improvements in ML Algorithms Implementation

There is a huge effort in the ML field to produce a general pipeline that covers all steps needed for algorithm implementation, but these efforts did not produce a robust pipeline to be used flexibly with different cases of data size, features size, data type, uncertainty in data, and complexity in data. Therefore, this paper aims to produce a general pipeline suitable for benchmarking and prediction in the building energy field.

Depending on the review of different pipelines resulting from previous trials, the proposed machine learning pipeline overcomes the drawbacks of each reviewed pipeline, explaining how to select and implement each machine learning approach on building energy benchmarking and prediction problem in a sufficient way. There are essential steps that must be found in the pipeline, and these steps will be described one by one. In addition, each ML algorithm has requirements to be selected as a solution tool for a real-life problem. Based on results from existing works and reviewed pipelines for different applications, a Pipeline is proposed to select and implement ML algorithms on real-life problems of the energy and buildings field.

2.1. Problem Identification and Formulation

In the beginning, the real-life problem is identified as building energy consumption benchmarking and prediction. From there, we began problem formulation which includes articulating the problem and converting it into an ML problem. Converting it to a machine learning problem requires us to identify features that should be found in the data to predict accurate output [10].

2.2. Data Collection, Analysis, and Preprocessing

Data have two elements: (1) a feature, which is an attribute that is used to help extract patterns and predict future answers, and (2) a label, which is an answer that is wanted from the model to predict. The data are collected by answering problem formulation questions, then converting answers to features' effect on output. After the problem is formulated, we need to ensure that the data are formulated correctly for the ML algorithm and cleaned up in a way that will maximize the performance of the model. Thus, the step of data collection, preparation, and preprocessing is very important [11].

This step includes the following. (1) Data collection and integration ensures that raw data are in one central, accessible place. The importance of this step appears when the results of the evaluation metric on training and test data are low because the learning algorithm did not have enough data to learn from. Thus, performance can be improved by using the data augmentation technique which increases the amount of data. (2) Data preprocessing involves transforming raw data into an understandable format and extracting important features from the data. (3) Data visualization entails several things including a programmatic analysis to give a quick sense of feature and label summaries, which is effectively helping understand the data [12].

There is a relation between the selection process of the appropriate ML algorithm and the nature of collected data. This relation depends on many factors: (1) data size, (2) features size, (3) data type (residential or non-residential data and time serious or not), (4) degree of uncertainty in data, and (5) degree of complexity in data.

The ANN is the most flexible algorithm in the popular ML algorithms. It has a high dynamic power that resulting from the flexibility in performance control by using different hyper-parameters values. The dynamics of ANN give this algorithm an advantage over other ML algorithms such as (1) handling huge data sizes in faster time with minimum computation power [13,14], (2) dealing with different data types by changing the type of ANN used (e.g., time serious data [13,15,16], annual commercial buildings' data [17,18], and residential buildings' data [19]), (3) it can overcome the complexity of data sets that have a lot of features because it gives high weights for important features during training, and it can be integrated with feature selection or feature extraction concepts [16,20,21], (4) it is integrated with other ML algorithms in different ways to increase performance [22], and (5) it can train on noisy data sets by changing the sensitivity of the trained model to changes of values [23] or use the Kalman filter [24] as preprocessing steps. The problems that keep ANN from an important role in the building energy field are that (1) ANN needs an experience to deal with the hyper-parameters tuning process to deliver the best performance [13,25], (2) the difficulty of identifying the most appropriate sample size that is suitable for real-life problems [25], and (3) decreasing prediction power with residential buildings' data [18,26].

The ability of ANN algorithms to handle big data is declared in different applications. In 2010, Dombaycı et al. utilized a total of 35,070 hourly temperature data to estimate the hourly energy consumption of a model house designed in Denizli, Turkey's Central Aegean Region, for selecting appropriate and efficient heating and cooling equipment, with 26,310 h used for training and 8760 h used for testing. (The ANN model was trained using heating energy consumption data from 2004 to 2007 and evaluated using heating energy consumption data from 2008.) The result states that energy consumption levels may be predicted with a high degree of accuracy and that the ANN is extremely successful with large data sets [13]. In 2015, Antanasijevi et al. developed a new approach for determining the accuracy of a GRNN (general regression neural network) model applied for the prediction of EC (energy consumption) and GHG intensity of energy consumption using historical data from 2004 to 2012 for a set of 26 European countries (EU Members). The result states that the GRNN GHG intensity model is more accurate than the MLR (multiple linear regression) and second-order and third-order non-linear MPR (multiple polynomial regression) models that were evaluated [14].

The importance of preprocessing steps declared in some previous papers, such as the complexity that results from increasing the number of features, was discussed in 2015 by Li et al. while improving short-term building hourly electricity consumption prediction. They utilized principal component analysis (PCA) as an automated approach to reduce the ML problem complexity, and they said that this technique was able to fulfill two goals (i.e., lowering ANN model complexity without compromising prediction accuracy) in only one automatic step [16]. In 2015, Platon, Dehkordi, and Martel used the same feature selection technique (PCA) to select the most significant features from all studied features (i.e., only 10 significant features were selected out of the 22 available features) to develop hourly electricity predictive models based on ANN [15]. In 2006, Karatasou, Santamouris, and Geros explained the ability to improve ANN performance by using statistical analysis (e.g., hypothesis testing and information criteria) as a preprocessing step before training to design an hourly building load predictor based on a feed-forward artificial neural network (FFANN) [21]. The concept of preprocessing steps for ANNs may depend on another ML concept that help in simplifying the process for complex problems. In 2014, Du et al. employed a clustering method to aid ANN algorithms in detecting abnormalities in air handling units, which are common in commercial buildings (e.g., fixed biases, drifting biases, and complete failure of the sensors and chilled water valve faults). For prior mistakes, the fault diagnosis tool for the HVAC system obtained good identification results [22].

The SVM algorithm is better than neural network algorithms, because of (1) the small number of parameters compared to ANN and genetic programming [27,28], (2) the SVM solution is unique and optimal because SVM can reach a global solution for problem [28,29], and (3) it can handle different types of data (e.g., time serious data [29], annual commercial buildings' data [18], and residential buildings' data [30]). On the other hand, the SVM algorithm cannot handle complex data that have too many features, so it is integrated with feature selection methods to decrease the number of problem dimension spaces by decreasing features [31]. In addition, it is not suitable for large data sets because the training process of SVM algorithms becomes very slow with a large amount of data, yet achieving good performance [28,29]. Sometimes, multi SVMs are used in parallel to reduce the computation time of large data [31,32].

The importance of preprocessing steps for SVM algorithms is greater than for ANN algorithms because it cannot handle the complex ML problems that have many features and nonlinear relations. Thus, in 2012, Zhao and Magoulès used correlation analysis for feature selection on complex data while assessing the energy demands of office buildings to reduce the number of features for suggested algorithms. By manually computing the linear correlation coefficients between characteristics and energy needs, the most significant features with significant correlations were chosen [31].

The GPR or GMM algorithms are the best ones to deal with noisy data or uncertainty in the data set. The reasons are as follows: (1) they overcome noisy measurements which come from sensors [33,34], (2) can extract complex patterns such as nonlinear and multivariate relations between features [33], (3) can be integrated with other ML algorithms as a preprocessing step to remove uncertainty in the data set [35], and (4) give very efficient and robust predictions results even if with a small size of data [33,34]. The main drawback of GPR or GMM algorithms is that they need high computation power and cost, especially with large data sizes [33].

Due to the ability of GPR and GMM algorithms to deal with complex ML problems and noisy data, the preprocessing steps do not have an essential role with these algorithms during their implementation in complex applications. In 2012, Heo and Zavala demonstrated that these algorithms could capture complicated behavior (i.e., nonlinearities, multivariable interactions, and time correlations). Furthermore, because they were created in a Bayesian environment, they have the potential to overcome problems of uncertainty, but require a lot of computing power to accomplish these findings in a short amount of time [33]. Moreover, the GPR and GMM algorithms can be used as a preprocessing step to filter noisy data. In 2012, using these algorithms, Heo, Choudhary, and Augenbroe detected uncertainty in buildings' measurements to improve modeling and retrofit performance while creating a scalable, probabilistic methodology [35].

The clustering algorithms are very powerful because (1) they can handle different types of data [36–38] and (2) propose a very powerful tool when integrated with prediction algorithms that increase prediction performance [36,39]. However, they have drawbacks such as (1) falling into the local minimum solution, especially the k-means algorithm, so it is recommended to iterate the clustering process to obtain the general solution; (2) they are affected by high data complexity, so it is necessary to apply feature extraction, feature selection, and PCA as a preprocessing step [36,40]; (3) they are affected by data uncertainty, so it is necessary to integrate the Kalman filter or GPR or GMM as a preprocessing step [24]; and (4) as the data size increases, the time of the iteration processes increases, too [38,40].

For increasing clustering algorithms' performance in obtaining global solutions, they can be integrated with preprocessing steps such as statistical analysis or feature selection in retrofit studies. In 2010, Gaitani et al. published an energy categorization tool for heating school buildings. Three steps were involved in the creation of the tool: (1) performing an extensive statistical analysis on the data, (2) applying PCA to select most significant features, and (3) using k-means clustering technique to classify. The conclusions declare that the proposed tool achieved very effective results because it used the two preprocessing methods with clustering for energy-saving techniques [40]. One of the big advantages of clustering algorithms is that they can be used as a preprocessing step for prediction models to increase performance. In 2017, Yang et al. demonstrated that combining the k-shape clustering technique with the SVR model as a feature extraction phase to produce new features from output clusters greatly improved the SVR model's hourly and weekly energy consumption forecasting accuracy [36].

2.3. ML Algorithm Selection

In general, determining which machine learning algorithm is the best is difficult. As a result, it is critical to thoroughly examine the type of accessible or gathered data as well as the application to select the most appropriate model. The ML algorithm selection step depends on many factors such as (1) application-type factors and (2) data factors. In this section, applications of four ML algorithms (i.e., ANN, SVM, GPR or GMM, and Clustering K-Means or K-Shape) are explained to represent advantages, drawbacks, and potential improvements for each algorithm to help ML users in the selection process for the most appropriate algorithm during implementation in the field. After reviewing ML applications, the ML users in the energy and buildings field can deduce that (1) ANNs are a strong tool for modeling and reliable prediction of building energy. They do, however, need a careful selection of network topology and fine tweaking of their many hyper-parameters for training. Because ANN suffers from a local minimum issue, the models' performance cannot be guaranteed. In addition, to obtain acceptable accuracy, ANN needs to be fed with a sufficient number of samples. Simple MLR models may be able to outperform them otherwise. As a result, ANN is best suited to engineers who are well-versed in deep learning and statistical modeling. We can also deduce that (2) SVM has been found to outperform ANN in load forecasting and can construct models from small data and (3) GPR is utilized for model training with uncertainty assessments among ML approaches and other black-box methods. Uncertainty and sensitivity analysis for various machine learning models have recently been presented and used. As a result, it is worthwhile to devote research resources to deploying these techniques for modeling construction under unclear data. Finally, we find that (4) in multi-dimensional energy assessment systems, k-means and k-shape are both highly efficient, with k-shape being used with time serious data and k-means being used with other data types. The popular applications of ML algorithms around the world in the field are explained in the following subsections and summarized in Figure 1.

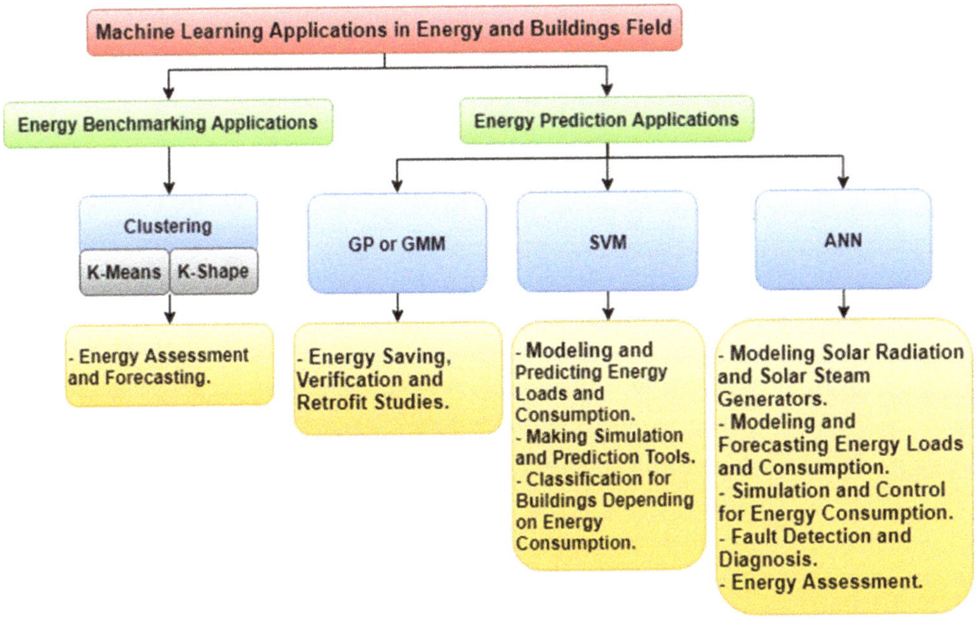

Figure 1. The popular applications for four types of ML algorithms around the world in the energy and buildings field.

2.3.1. Applications of ANN Algorithm

ANN algorithm represents a very powerful tool in the energy and buildings field because it can simulate the human brain by using nodes, weights, and layers to store information in parallel paths and it is extracted when necessary in a parallel way, too. This section demonstrates the ability of ANNs to be used in different applications of the energy and buildings field, with a description for implementation methods and contributions.

Modeling Solar Radiation and Solar Steam Generators

Kalogirou, in 1998, demonstrated several ANN applications in the field of solar energy. The author utilized artificial neural networks (ANNs) to predict solar radiation and a solar steam generator at varied incidence angles. ANN uses climatic parameters to estimate hourly solar irradiance in solar radiation modeling. ANN can forecast the collector intercept factor (i.e., the ratio of the energy absorbed by the receiver to the energy incident on the concentrator aperture) for solar steam generator design, as well as the radiation profile and the heat-up temperature response [41].

Modeling and Forecasting Energy Loads and Consumption

ANNs also take the place in energy simulation systems because they achieve fast computation time and performance in many applications. Olofsson et al. developed a long-term prediction and performance evaluation tool based on artificial neural networks (ANN) in 2001, using data from two to five weeks for six building families in Sweden built in 1970. The authors utilized the PCA approach to reduce the number of characteristics to only four (e.g., construction year, number of floors, framework, floor area, number of inhabitants, and ventilation system). The tool was created by using short-term data to evaluate performed retrofits and present conditions for improving the exited buildings, and made a long-term prediction for building energy consumption [20]. Ascione et al., in 2017, studied the ability to create a building energy prediction tool with low computational power and high accuracy. The tool was constructed using data from office buildings erected in southern Italy between 1920 and 1970. Based on the ANN algorithm, the authors presented two concepts of prediction tools: (1) used the existing data as it is, and (2) used the existing data but in the presence of energy retrofit measures. The proposed ANNs were optimized by "Simulation-based Large-scale sensitivity/uncertainty Analysis of Building Energy performance" (SLABE). The performances of the networks were estimated by using the distributions of the relative error to compare ANNs' outputs with EnergyPlus program targets. The conclusion declares that the developed ANNs can replace standard building performance simulation tools, thereby reducing computational effort and time [42]. Beccali et al. studied EU non-residential building energy consumption in 2017 to develop an energy evaluation tool based on two artificial neural networks (ANNs), the first of which was used to forecast actual energy consumption and the second of which was used to assess economic indicators. The authors used 151 existent buildings in four locations of southern Italy to assess the two ANNs. The conclusion states that the decision support tool based on ANNs was able to forecast the energy performance of buildings quickly and accurately and that it was used to pick energy retrofit alternatives that can be implemented [43].

The ANNs' flexibility is also declared in different applications when integrated with other optimization techniques to improve overall performance. Paudel et al. in 2014, integrated a pseudo-dynamic technique with an ANN model to make a daily short-term prediction for building heating demand. Because the hidden information in heating demand cannot be retrieved from climatic data by using ANN alone, the pseudo-dynamic approach improved the overall performance of the ANN model by aspects of operational heating power characteristics. The algorithm is used in the construction of French institutions. The created dynamic model is resilient, according to the conclusion, and may be utilized by energy service companies (ESCOs) in heat production dynamic control systems [44]. In 2017, Ascione et al. presented a detailed analysis and forecasting method based on ANN for cooling load of institutional building. For two years, data were collected from three institutional buildings. Due to the nature of vacation times and university timetables, the research reveals a large variance in daily cooling loads energy consumption. The authors proposed dividing the data into groups based on vacation times and university timetables to solve the problem of variation and examine it. The conclusion states that by adding categories' numbers as a new input feature to the ANN algorithm, it was able to improve predicting accuracy. Furthermore, by utilizing the Bayesian regularization approach for the hyper-parameters automated tuning process, the performance of ANN can be most effective and rapid in computation time [45].

We can analyze the performance of the ANN algorithm declared in some previous papers to highlight the advantages and drawbacks by comparing it with other ML algorithms in the energy and buildings field. In 2015, Platon, Dehkordi, and Martel presented hourly electricity predictive models based on ANN and case-based reasoning (CBR) for an institutional building. The measured data from a Canadian institutional facility included elements, such as weather information, that are relevant to the building's operation. To forecast power usage with a horizon of 1 to 6 h, the authors utilized principal component analysis to identify the most important characteristics (i.e., only 10 significant features were chosen out of 22 available features). The models' prediction abilities were evaluated, and the ANN models regularly outperformed the CBR model, according to the results. Both the CBR and ANN models, on the other hand, had an error that was well within the ASHRAE limits. To improve the CBR model's performance, different approaches were tried: (1) varying the case similarity criterion and the number of previous instances used for prediction and (2) using automated optimization techniques on values' weight. However, none of these techniques had a substantial impact on the CBR models' performance [15]. Edwards, New, and Parker, in 2012, sought to reduce difficulties such as a large number of features in the building characterization and prevent the problem of energy consumption in the predesign stage. The scientists used sensor-collected energy usage data to conduct statistical analysis using several machine learning methods (e.g., feed-forward neural network, support vector regression, least squares support vector machine, a hierarchical mixture of experts, and fuzzy c-means with feed-forward neural networks). To forecast next hour energy usage, researchers compared several machine learning algorithms on two types of data: (1) data on commercial building consumption gathered hourly and (2) data on residential building consumption collected every 15 min. According to the findings of this comparison, ANN-based techniques perform better on commercial structures. However, results show that these methods perform poorly on residential data and that least squares support vector machines perform best on both, but with high computation costs [18]. Kialashaki and Reisel, in 2013, created a hybrid method using artificial neural networks (ANN) and multiple linear regression techniques to forecast future energy consumption for residential buildings in the United States under various input scenarios (e.g., dwelling size, number of occupants, the efficiency of heating equipment and energy intensity). The authors describe how ANN's effectiveness varies in residential structures in the United States, especially with test data. As the ANN model prediction is dependent on the cumulative trends of the various parameters, the reason for the variation in forecast energy was the fluctuation induced by the economic recession [19].

Simulation and Control for Energy Consumption

The energy consumption for buildings can be enhanced and controlled easily by using the ANN algorithm because it has the ability to deal with nonlinear equations in some applications. In 2015, Huang, Chen, and Hu proposed predictive control for an HVAC system to forecast an interior temperature by taking into account nonlinear building thermal dynamics (e.g., interaction between locations, noise in sensors, and delay time). Energy input from mechanical cooling, ventilation, weather, and convective heat transfers for thermal coupling between locations are all features of the ANN input. The suggested ANN model incorporates the thermal interaction between zones, resulting in more accurate prediction results than a single zone model, according to the conclusion. This management approach resulted in a high level of building energy consumption control [23]. In 2016, Benedetti et al. presented an automatic tool based on ANN to control building energy consumption and investigated the effect of the collected data period on the automatic utilization of such tools where a large amount of data is not always available in the real world, so the minimum and maximum period of required data were identified to achieve reliable results. To determine the optimal ANN design for an energy consumption management tool, the authors used three alternative ANN architectures. Furthermore, because a large quantity of data is not always present in practice, a method is presented for determining the minimum time of data collection required to achieve accurate findings and the maximum period of usefulness [46]. In 2017, Ahn, Cho, and Chung presented a hybrid control approach on mass and temperature for supply air of heating system to minimize energy consumption. To understand the nonlinear relations between features and forecast or assess precise thermal, the suggested technique uses a mix of fuzzy inference systems and ANN. To assess supply air conditions for a heating season, the suggested technique was compared to a basic thermostat on/off controller, and it was discovered that the ANN controller can reduce energy usage when compared to a simple thermostat on/off controller [47].

Fault Detection and Diagnosis

Time consumption problems during energy assessment and retrofit studies for buildings vanish in some studies. Kalogirou et al. proposed a fault diagnostic prediction system in 2008 that used temperature readings to identify problems in solar water heater components and forecast mistakes in collectors or pipe insulation. There were four elements to the problem diagnosis system: (1) a data acquisition system measured temperatures in four locations of the solar water heater system and the mean value for a storage tank; (2) a prediction module based on an artificial neural network (ANN) that was trained with fault-free system values obtained from a TRNSYS under the same meteorological conditions (e.g., Nicosia, Cyprus, and Paris, France), (3) the residual calculator takes measurement data from the data collection system as well as error-free predictions from the prediction module, and (4) the diagnosis module detects a variety of defects, including collector faults and insulation failures in the pipes linking the collection to the storage tank [48].

Energy Assessment

In 2013, Hong et al. studied the energy performance of schools (from 2008 to 2011) to create energy evaluations by combining statistical analysis with artificial neural networks (ANN) to evaluate the influence of each feature on energy and the relationship between them. About 7700 schools were utilized in a rapid statistical study, and 465 schools were investigated in depth using ANN to find variables that influenced school energy usage patterns. The results declared that the non-domestic buildings must be re-classified because of different reasons: (1) changes in the energy use pattern and (2) differences in energy performance between primary and secondary schools such as a gradual increase in electricity consumption and a decrease in heating consumption in both. By comparing simulation and engineering calculations, the authors noted the ability of ANN in energy assessment and the limitation in prediction [49]. Buratti, Barbanera, and Palladino developed a verification tool based on ANN in 2014 to forecast energy consumption and assess building performance by comparing it to energy certificates. The Umbria Region (central Italy) acquired around 6500 energy certificates (2700 of which were self-declarations). To train the ANN, the authors utilized only right certificates recognized by comparing them to energy standards, and they created a new index called the neural energy performance index to describe the degree of accuracy and to identify the certificate's precise control needs (NEPI) [50].

2.3.2. Applications of SVM Algorithm

The SVM algorithm represents the best alternative solution for the ANN algorithm in many applications of the energy and buildings field. SVMs have a low number of hyper-parameters compared to ANN models, so they are easier to control and can be trained with small data sizes.

Modeling and Predicting Energy Loads and Consumption

The power and drawbacks of the SVM algorithm appeared in many applications of the field by comparing it with ANN models to solve the same problems. By using the SVM method for hourly cooling load forecast of an office building in Guangzhou, China, Li et al. demonstrated in 2009 that it is extremely successful, even with small data sets. The findings were compared to those of backpropagation ANN to indicate that SVM outperformed ANN in terms of accuracy and global solution. The input features were (1) outdoor dry bulb temperature of the past 2 h and (2) solar radiation intensity of the past 1 h. The result states that the SVM algorithm performed as well as the ANN method in terms of speed and accuracy, but with fewer data samples [28]. Using the least square support vector machine, Xuemei et al. increased the time efficiency required for hourly cooling load forecast in 2009 (LSSVM). The authors compared the proposed approach to backpropagation ANNs to assess its performance. In the end, LSSVM outperformed backpropagation ANN in terms of accuracy and global solution, especially when the available training set is restricted. As a result, LSSVM might be a viable option for predicting the cooling demand in a building [29].

After proofing the ability of SVM to replace ANN in different applications of the same field, there are different papers applied the algorithm with some adjustments to overcome drawbacks such as high time consumption when used with large data size. Hai Xiang Zhao et al. in 2009, studied the ability of SVM with the Gaussian kernel algorithm to deal with large time series datasets and reduce the training time of predicting energy models by using a concept of parallel SVM algorithms. Results showed very good performance in the prediction of energy consumption in multiple buildings based on large time series datasets [32]. Zhao and Magoulès, in 2012, studied the ability to reduce time consumption in SVM training with large data size by using radial and polynomial functions as a kernel for parallel SVM algorithms to predict the energy consumption of office buildings. The algorithm feature selection is implemented on data by using correlation analysis for features. Using correlation analysis for features, the algorithm feature selection is implemented on data. To compute the energy demands, the authors utilized simulated data from EnergyPlus software and manually selected features by computing correlation coefficients to reduce the number of features for the proposed algorithms [31].

Making Simulation and Prediction Tools

SVM algorithms can also outperform the ANN algorithm in dealing with residential buildings data for many cases such as energy prediction and creating tools. Jain et al., in 2014, used a support vector regression (SVR) algorithm with sensor measurements from residential buildings to make energy predictions. The inputs feature during training were (e.g., weather, time of day, and previous energy consumption) from multi-family residential building data in New York City. The authors mentioned a paucity of research applying multi-family residential buildings. Thus, he expanded the study-to-study algorithm limitations by examining different time steps (i.e., 10 min, daily, and hourly) and different spatial categories (i.e., by unit, by floor, and whole building). The conclusion declares that the SVR could be used in energy prediction for residential buildings and the best prediction results occurred at floor level in hourly intervals [30]. In 2008, Lai, Magoulès, and Lherminier utilized SVM to develop a simple and rapid method for predicting the electric energy consumption of residential buildings. The data include daily electricity usage for a year and three months, as well as climatic data such as temperatures and humidity. For the learning stage, the authors utilized a year and two months, and for the prediction step, the authors used the last month. The findings demonstrate that the model has high performance and that the SVM tool may be utilized to conduct predictive modeling [51].

Classification for Buildings Depending on Energy Consumption

The SVM algorithm is very flexible to be integrated into existing systems on building energy management. In 2010, Li, Bowers, and Schnier developed a daily power consumption management system for buildings based on detecting abnormal energy behavior and providing the capacity to handle problems in real time to enable prediction and detection of abnormal energy usage. The system consisted of the following steps: (1) outliers' detection in real time to identify abnormal energy use and delete it from further analysis, and (2) classifying based on the SVM-predicted daily electricity profile. The suggested system was computationally efficient and resilient enough to be incorporated into current building energy management and alarm systems [52].

2.3.3. Applications of GPR or GMM Algorithm
Energy Saving Verification and Retrofit Studies

Although the Gaussian-based algorithms need high computation power resources, they have a lot of advantages declared through implementation in some complicated applications which make them used in the field. In 2012, Heo and Zavala investigated the possibility of the GPR model to substitute a linear regression approach in energy savings, uncertainty measurements, and verification problems since it is highly powerful in prediction, particularly with noisy data. The conclusion asserts that generalized linear models (GPR models) can represent complicated behavior (i.e., nonlinearities, multivariable interactions, and time correlations). Furthermore, because they were created in a Bayesian environment, they can overcome difficulties of uncertainty [33].

These solution algorithms are best in the case of noisy data or probabilities and retrofit studies to help decision makers in taking steps in improving countries. Furthermore, in 2012, Heo, Choudhary, and Augenbroe presented a scalable, probabilistic methodology for energy modeling based on Bayesian calibration (the same base for Gaussian models) to improve modeling by detecting uncertainty in buildings models and aid in studying the probability of building energy consumption improvements and retrofit performance. The suggested technique, according to the conclusion, may accurately assess energy retrofit choices and promote risk-aware decision-making by clearly inspecting risks associated with each retrofit option [35].

The GPR and GMM algorithms can be integrated with other models to improve performance even if the collected data are limited. In 2014 Burkhart, Heo, and Zavala utilized a GPR with a Monte Carlo expectation maximization (MCEM) model to cope with noisy data from sensors (e.g., weather, occupancy) and investigated the impact of the method on the quantity of necessary data from sensors during measurement and verification (M&V) stages. The GPR-MCEM model, according to the result, reached robust prediction levels when compared to conventional GPR alone, and may be utilized as a mechanism to decrease data collection and sensor installation costs in M&V processes since it provides high performance with fewer data [34].

2.3.4. Applications of Clustering Algorithms (K-Means and K-Shape)
Energy Assessment and Forecasting

The benchmarking process is very helpful in building energy assessment applications, especially when integrated with other algorithms to create energy assessment techniques. In 2007, Santamouris et al. developed an intelligent technique to cluster school buildings as the first step in energy assessment procedures. Then, the output clusters used in the energy performance studies specified the buildings' rating and environmental impact of each cluster. The energy rating of the school buildings gives detailed information on their energy consumption and efficiency in comparison to other buildings of a similar kind, allowing for better intervention planning to enhance their energy performance. The authors created the technique in three steps: (1) energy consumption data were collected from 320 schools in Greece, (2) fuzzy clustering techniques were used to make the energy rating scheme, and (3) 10 schools were selected and detailed analysis was performed for energy efficiency, performance, and environmental impacts. The conclusion declares the ability of the used technique to identify and rate the existing school buildings and studied the potential for energy and environmental improvements [39]. Gaitani et al., in 2010, presented an energy classification tool for school buildings' heating based on a k-means clustering technique with PCA to help decision makers in the schools rating process and study probabilities of energy savings. The data used consisted of 1100 cases from secondary education school buildings in Greece, which represented 33% of the total secondary school sector, and included information such as energy consumption for space heating and lighting, building area, number of students and professors, a boiler installed power, building manufacturing year, and operation schedule. The tool was created in three steps: (1) an extensive statistical analysis on the data was performed, (2) PCA was applied to select the most significant features, and (3) a k-means clustering technique was used to classify. The results state that the categorization may be used to aid decision makers' energy-saving strategies [40]. In 2017, Yang et al. proposed an energy clustering method based on the k-shape algorithm for time series data, which can recognize patterns in time series data and categorize them using multi-dimensional space. The clustering was performed on data from 10 institutional buildings' hourly and weekly energy usage using the k-shape method to find form patterns in time series data, which increased the accuracy of forecasting models. The conclusion declared that the proposed method could detect building energy usage patterns in different time intervals effectively and also proved that the forecasting accuracy of the SVR model is significantly improved by integrating the clustering method with the SVR model [36].

These clustering algorithms also prove efficiency as an alternative solution for software such as the Energy Star program. In 2014, Gao and Malkawi proposed a benchmarking technique based on the smart clustering concept, which classifies buildings' energy based on all features that have a relationship with energy consumption and groups buildings with the most similarity of features into one cluster, implying that the problem of classifying is multi-dimensional. The proposed methodology contains four steps: (1) data collection, (2) feature identification and selection, (3) selection for clustering algorithm depending on collected data, and (4) buildings' benchmarking concerning cluster group and centroid. The findings were compared to the Energy Star approach to show that the suggested strategy can give a more thorough approach to benchmarking, particularly with multi-dimensional challenges, inspiring a fresh viewpoint on building energy performance benchmarking [37].

The complexity of ML problems can be handled by the clustering algorithms, which convert the chaos data to more homogenous ones in simple iterative steps. Arambula Lara et al. in 2015, studied the European policy of energy saving and the Commission Delegated Regulation (EU) 244/2012, which gave recommendations for some reference buildings to make a compromise cost from expected improvements. The solution was found in the k-means clustering approach, which split huge data into tiny and homogeneous groups based on building characteristics' similarity, decreasing the complexity of energy optimization and retrofits by reducing school buildings' stock homogeneously. The data came from a sample of roughly 60 schools in the region of Treviso in northern Italy, collected between 2011 and 2012. The conclusion declares that this method could identify a small number of parameters to assess the energy consumption for air heating and hot water production [38].

2.4. Model Training, Validation, and Tuning

This is an iterative process during the conversion of a solution that can be performed many different times. Initially, upon training, the model will not achieve the results that are expected. Thus, the tuning process is very important to evaluate model performance under different values of hyper-parameters. During training, the machine learning algorithm updates a set of numbers known as parameters or weights. The goal is to update model parameters in the global solution direction which makes the predicted output as close as possible to the true output (as seen in the data). This cannot be achieved in one iteration, because the model has not yet learned; it watches the weights and outputs from previous iterations and shifts the weights to a direction that lowers the error in the generated output. If the error in the output gradually decreases with each successive iteration, the model is said to converge, and the training is considered successful. If, on the other hand, the errors either increase or change randomly between iterations, the hyper-parameters of the model need to be tuned [12].

The most important thing in model performance is overfitting and underfitting. The underfitting problem means that the model performance is very low on the training data, and thus, the training model is unable to represent the data correctly. In underfitting problems, the model could be very simple (the problem cannot be formulated well with enough features) to produce accurate outputs well, because of the inability to extract patterns or relationships between input and output features for data. To overcome the underfitting problem, there are different techniques: (1) reformulate the real-life problem by adding more effective features, (2) choose suitable preprocessing methods to solve data drawbacks of missing or outlier values, and (3) decrease or change amount or type of model regularization techniques such as dropout. The problem of overfitting is present when the model performance is very high on training data but low on the validation or test data. The reason for the overfitting problem is the inability of the model to attain the global solution of the problem to cover all data sets. With low performance for unseen data, it makes sense to use fewer feature combinations and increase the amount of regularization [53].

There are several techniques to overcome overfitting and maximizing generalization. The most popular one is simple Hold-Out Validation. The simple hold-out technique depends on splitting data into multiple sets for training, validating, and testing models. Training data, which include both features and labels, feed into the model. The model is then used to make predictions over the validation data set, which checks performance to tune and change the model's weights. Then, test data that only include features are used to produce the labels. The performance of the model with the test data set is what we can reasonably expect to see in real life [13,54,55].

The problem of the hyper-parameters tuning process is mainly related to the ML algorithm type. For the ANN algorithm, many papers discussed this problem during the implementation of different real-life problems and recommended using an automatic technique in the tuning for a large number of ANN hyper-parameters because the ANN model has a lot of hyper-parameters.

González and Zamarreño, in 2005, studied the effect of hyper-parameters such as the number of neurons per layer and data size on the performance of ANN while creating an algorithm for short-term building load prediction. The authors mentioned difficulties in reaching the global solution because it related to large numbers of ANN hyper-parameters values [25]. Dombaycı et al. in 2010, studied the number of neurons per layer only as a hyper-parameter for an ANN model that was developed to make an hourly heating energy prediction in the design stage for a building to help in selecting appropriate and efficient heating and cooling equipment. The authors explained the complexity of tuning ANN hyper-parameters by using manual methods because of their large numbers [13]. One of the trials to overcome the problem of a large number of hyper-parameters was conducted in 2015 by Li et al., who used particle swarm optimization technique in automatic ANN hyper-parameters tuning while improving the short-term building hourly electricity consumption prediction and compared this method and simple ANN with manual tuning for hyper-parameters. The authors concluded that the automatic tuning process has a shorter training time and higher performance than the manual method and the hybrid genetic algorithm model [16]. In 2017, Ascione et al. presented a solution for the same problem using the Bayesian regularization technique for hyper-parameters' automatic tuning while making a detailed analysis and forecasting method based on ANN for the cooling load of institutional buildings, and mentioned that the performance of ANN is the most effective and quick in computing time by using this tuning technique [45].

Otherwise, the SVM algorithms are easily tuned and manually optimized. These advantages appeared in many applications such as in 2009 when Zhijian Hou et al. studied the ability to replace huge numbers of trainable parameters for ANN by using radial function as a kernel for an SVM algorithm in an HVAC system energy prediction in Nanzhou. The paper proved that the algorithm has fewer parameters to tune compared with ANN and is better than the ANN algorithm in forecasting [27].

2.5. Model Evaluation

The model evaluation is performing using test data to make sure that the required goal is achieved and to overcome the problem of over-fitting and under-fitting for the trained model. If the trained model does not meet the required goal, it increases the required time to re-validate the model and achieve goal. In this step, the feature engineer takes the role to study data and features and find ways to improve the model, and the way that it is produced. Once the retraining happens and the required goal is achieved, the model is deployed to perform the best possible predictions on the unknown data to begin evaluating how the model responds in a non-training environment.

To evaluate the machine learning model, we need to know the type of ML problems, classification (such as benchmarking), or regression (such as prediction) problems. The type of machine learning problem will influence the type of metric used to evaluate the model. We can start by looking at classification problems metrics. There are different types of metrics to evaluate models: (1) accuracy, (2) precision, (3) recall, (4) F1 score (2*(recall*precision)/(recall + precision)), and (5) area under the curve-receiver operator curve (AUC-ROC). To implement this metric method on the trained model, the model predictions and the known target values are sent to the confusion matrix. A confusion matrix is the building block for running these types of model evaluations for classification problems. Then, the predictions are returned, compared with the values of ground truth. Finally, the evaluation metric between predicted values and ground truth values is computed [56].

It is recommended in classification problems to use F1 as an evaluation metric because the F1 score combines precision and recall together to give one number to quantify the overall performance of a particular ML algorithm. In addition, the F1 score should be used when the dataset has a class imbalance but we want to preserve the equality between precision and recall.

In regression problems, there are other common metrics that we can use to evaluate the model: (1) mean squared error and (2) R-squared. Mean squared error is very commonly used. The difference between the prediction and actual value is calculated, that difference is squared, and then all the squared differences for all the observations are summed up [16].

The other metric type is R-squared, which explains the fraction of variance accounted for by the model. It is like a percentage, reporting a number from 0 to 1. When R-squared is close to 1, this usually indicates that a lot of the variabilities in the data can be explained by the model itself. The threshold for a good R-squared value depends on your machine learning problem. In some machine learning problems, it is very difficult to achieve a high R-squared value. The high value of R-squared does not always represent strong model performance because R-squared is always increasing when more variables are added to the model, which sometimes leads to overfitting [13]. To counter this potential issue, there is another metric for model performance called the Adjusted R-squared value. The Adjusted R-squared has already taken care of the added effect for additional variables and it only increases when the added variables have significant effects on the prediction. The adjusted R-squared adjusts the final value based on two factors: (1) the number of features and (2) the number of data points in the data. A recommendation, therefore, is to look at both R-squared and Adjusted R-squared. This will ensure that the model is performing well but also that there is not too much overfitting.

In the building energy prediction field, it is preferred to evaluate ANN using mean absolute percentage error (MAPE) as a performance metric during model training. It is used in different applications in the same field and is proved to be very effective in the examination of model quality during the prediction process [25].

The most recommended evaluation technique is the cross-validation method that was used in 2006 by Karatasou, Santamouris, and Geros. They evaluated the hourly buildings load predictor based on feed-forward artificial neural network (FFANN) by splitting the data into many packages and looping them in the training process (i.e., each iteration in the training process carried out by using one of the packages as test data and others as training data to cover all data samples without overfitting problems). In addition, the authors discussed the cross-validation technique effect during training on the result of prediction and modeling and recommended this technique in such applications to achieve more robust models. The authors also discussed, the importance of attaining a more robust model by using different types of data sets in the evaluation process (i.e., the model performance was evaluated using two different data sets: (1) energy prediction shootout I contest and (2) an office building in Athens) [21]. Furthermore, one of the evaluation techniques was used by Dombaycı et al. in 2010 while developing an hourly heating energy prediction model based on ANN for building. The total data of 35,070 h were split into two packages to train and test the model: (1) the data from 2004 to 2007 (i.e., 26,310 data sample) used during model training, and (2) the data of the year 2008 (i.e., 87,60 data sample) used in model testing or evaluating. The authors mentioned the importance of using test or unseen data to improve model performance in real life [13].

2.6. Model Verification

The verification of machine learning models' robustness refers to checking models deployed on a real-life problem to ensure that it adheres to these specifications and achieves the target for a long run. A variety of machine learning models are also assessed according to how robust they are proven to be. This step must be performed frequently to make sure that the system is still working in high performance.

The evaluation step represents the base for verification steps during model deployment. The more models are robust in the evaluation step, the easier the verification step, and results in real life will be better. The robustness or verification for any model is examined firstly during the evaluation step by using unseen data or new data that differed from training data. Different types of data packages in many applications are used in the ML model evaluation step, such as in the 2006 work Karatasou, Santamouris, and Geros, who evaluated the FFANN models by using two different data sets: (1) energy prediction shootout I contest, and (2) an office building in Athens. In addition, they trained models on different time steps to identify limitations for models and create a robust hourly buildings load predictor tool so that the FFANN can be deployed on different data sets and used for a long run [21]. Moreover, in 2015, Li et al. collected hourly data from two resources: (1) energy prediction shootout contest I, and (2) a campus building in east China; meaning that the data were collected from different locations all over the world to ensure the reliability and robustness of the model [16].

3. Discussion

As shown in Figure 2, the generic ML covers all required steps to use and deploy the ML algorithms (i.e., ANN, SVM, GPR or GMM, and k-mean or k-shape clustering) in the energy and buildings field. The pipeline starts with problem identification (i.e., identify application type and specify the benchmarking and prediction problems). Then, this real-life problem must be converted to an ML problem in the problem formulation step by identifying the related features. After that, the data scientists start in collecting the data depending on the related features identified in the previous step and make some statistical analysis and visualization to study the nature of collected data and their distribution (this step is very important to help data scientists in choosing the appropriate preprocessing techniques).

Thereby, the data preprocessing step starts with answering some questions: (1) are there too many features? If there are too many features, the features must be decreased by feature selection (i.e., keeping only the most significant features that have a high effect on the studied problem). If there are not too many features, the second question is (2) are there too few features? If the collected data have a small number of features that have nonlinear or deep interaction relationships, the data scientist must employ some feature extraction techniques to increase the number of features and help the algorithm reach for a global problem solution during training. The third question concerns (3) noisy data. If the data contain noise, the filtration must be carried out by Gaussian based models or the Kalman filter. The fourth question concerns (4) time series data. To identify the noise filter type, this question must be answered to select between Gaussian-based models or the Kalman filter. The final step in preprocessing steps is solving problems of outliers and missing values.

Then, the most appropriate ML model must be selected. The selection depends on the ML problem type (i.e., benchmarking or prediction) that is identified in the first step in the proposed pipeline. In addition, some questions must be answered to identify the algorithm type. The first question concerns (1) time series data. If the data are time series, the k-shape clustering algorithm is the best selection for benchmarking. If not, the k-means clustering algorithm is better. The second question concerns (2) very big data. If the data are very big, the ANN algorithm is the best solution for prediction problems. If not, the next question concerns (3) complex systems. If the data have nonlinear or deep interaction relationships between features, the GPR or GMM are the best solution models for prediction problems. If not, the SVM model is better. After that, the labels from clustering is appended to the data in benchmarking to analyze the result, and the data must be normalized before training in prediction problems.

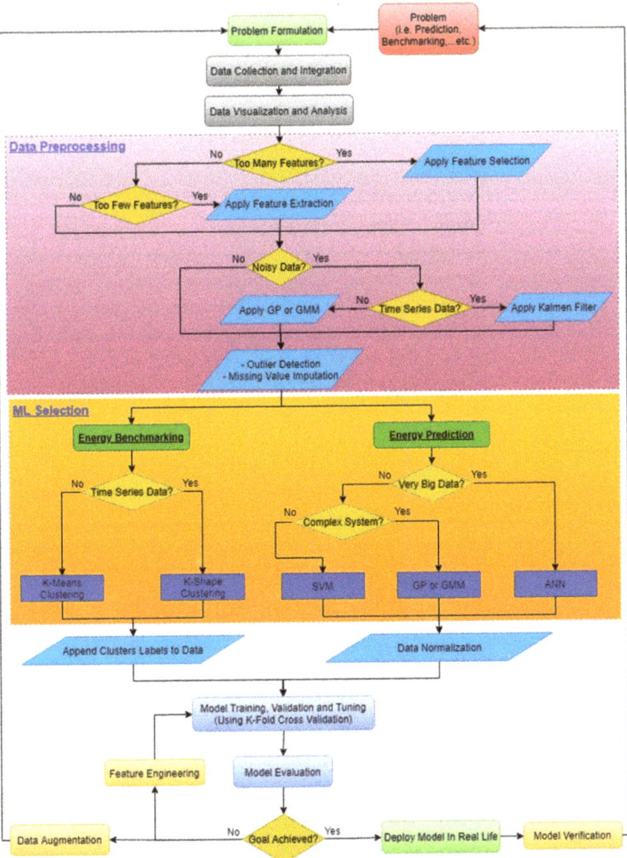

Figure 2. Building energy prediction and benchmarking pipeline.

Then, the model is trained, validated, tuned, and evaluated by using the cross-validation technique to create a robust model. Finally, if the evaluation results achieve the model goal, the model is deployed in real-life applications and make verification processes regularly. If the results are bad, the feature engineering must be conducted to increase or decrease features more and more (i.e., feature selection or feature extraction) and return to the training step again, or the collected data are not enough and must be increased by a return to the problem formulation again.

The final ML pipeline is very important for ML users in the energy and buildings field because it abbreviates a high level of experience in one pipeline to avoid the time consumption from new ML users to learn the potential of each ML algorithm. For more demonstration, we explain some previous work in the field by using the proposed pipeline (in Tables 1 and 2) to prove that the pipeline can be used as a reference for implementing the ML concept in the right way and achieving high performance:

(1) In 2006, Karatasou, Santamouris, and Geros designed an hourly buildings load prediction tool based on a feed-forward artificial neural network (FFANN). By comparing between paper steps and the proposed pipeline, it is found that the authors did not mention any preprocessing steps except statistical analysis. They stated that the data did not have any noise, removed the missing values, and normalized the data. Thus, they did not take the full benefits of statistical analysis to study the data nature, and there are some wrong prediction peaks due to ignoring the outliers' effect in the

preprocessing data step. Because of the large data size and since the ML type is prediction, the selected algorithm was ANN, and they implemented the cross-validation technique to create a robust model. In addition, the ANN algorithm was evaluated with two different data sets to ensure robustness, but it is not enough because the evaluation would be better if performed on the same model structures with different data sets but with the same input features to increase reliability and robustness [21].

(2) Dombaycı et al., in 2010, developed an hourly heating energy prediction model based on ANN to estimate energy in the design stage. The authors did not mention any preprocessing steps, just normalization, because the user data were calculated, so the probability of containing noise, missing values, and outliers is very small (this does not have the same worth of actual data). The ANN was used because the data are big and the ML problem concerns prediction. The data were split to train and test sets, but this was not enough because the trained model could be more robust if the cross-validation technique was used in training and evaluation steps [13].

(3) Mena et al., in 2014, developed and assessed a short-term predictive ANN model of electricity demand. The authors manually reduced the number of features because the data had a high number of features. Although the authors mentioned the outliers and noise in the data, they did not apply any type of analysis to solve these two problems in the data. In addition, the missing values in the data are kept as is and the authors depended on a manual method in splitting the data to skip missing values, which means the splitting blocks are imbalanced. Thus, the efforts made in the training and evaluation steps to create a robust model are useless because the preprocessing steps are not well performed, so the results from the model have a relatively high mean error [57].

(4) In 2015, Li et al. improved the short-term hourly electricity consumption prediction of a building. The authors mentioned a large number of features, so they used an automatic method of reducing features (PCA). However, the authors did not mention anything about the missing values and outliers in the data. Because of the large data size and since the ML type is prediction, the selected algorithm was ANN. The automatic tuning gives high prediction results, but it needs to integrate with cross-validation techniques to ensure the robustness and reliability of the model [16].

(5) In 2017, Yang et al. proposed an energy clustering and prediction method based on k-shape and SVM algorithms for time series data. The authors mentioned the noise in the data but did not mention solving it. In addition, there was no mention of any technique to solve the problem of outliers and missing values. The data size is relatively high to be used in SVM algorithms (the authors did not take into consideration the data size when selecting the algorithm), and the authors extracted features to decrease complexity and effort during model training. Due to the huge data size, it is recommended to used parallel SVM to reduce time or replace it directly with ANN [36].

(6) Heo and Zavala, in 2012, used the GPR model in energy savings and uncertainty measurements and verification problems. The authors did not use any feature extraction concept, although they mentioned a high degree of complexity in the data due to noise and nonlinear relationships. Moreover, they did not mention any technique to detect and solve the outliers and missing values problems. The data size is relatively large and since the authors did not mention time consumption in training, it may be too large. Thus, it will be better to use the Gaussian model to remove noise only and complete the prediction by ANN or use ANN directly for all problems [33].

(7) In 2014, Gao and Malkawi proposed a benchmarking technique for building energy based on the k-means concept. The authors used the features selection technique due to the high number of features. The data contain outliers, but the authors did not mention the technique to solve this. In addition, the imputation technique for missing values was not declared well, which greatly affected the k-means solution (the k-means has a high probability of falling into local minimum) [37].

Table 1. Review of some previous papers comparing them with the proposed ML pipeline.

| Preprocessing Questions and Actions | | | | | | | Model Selection Questions and Actions | | | | | | Model Creation | | |
Too Many Features? or Too Few Features?	Extracted or Selected Features	Noisy Data?	Time Serious Data?	Kalman or Gaussian Filters	Outliers' Values	Missing Values	Benchmarking or Prediction?	Time Serious Data?	Very Big Data?	Complex System?	Selected Algorithm?	Append Clusters Labels to Data? or Normalize Data?	Training, Validation and Tuning	Evaluation	
Not asked	—	×	√	—	—	removed	prediction	√	√	Not asked	ANN	normalized	Applied cross-validation on two data sets	Used two data sets and tried different samples steps	[21]
Not asked	—	×	√	—	—	—	prediction	√	√	Not asked	ANN	normalized	Split data to train and test sets	Used test data	[13]
Too many features	Selected features using correlation between features	√	√	—	Keep as it	Keep as it	prediction	√	√	√	ANN	normalized	Split data to train, validate, and test sets with different samples steps	Used test data with different samples steps	[57]
Too many features	Selected features using PCA	×	√	—	—	—	prediction	√	√	Not asked	ANN	normalized	Split data to train and test with automatic tuning (PSO)	Used test data	[16]
Too many features	Extracted features using k-shape clustering	√	√	—	Filtered	Imputed	Benchmarking and prediction	√	Not asked	Not asked	k-shape and SVM	normalized and append cluster labels	Split data to train and test and applied cross-validation	Used test data with different samples steps	[36]

Table 1. *Cont.*

	Preprocessing Questions and Actions							Model Selection Questions and Actions						Model Creation		
Too Many Features? or Too Few Features?	Extracted or Selected Features	Noisy Data?	Time Serious Data?	Kalman or Gaussian Filters	Outliers' Values	Missing Values	Benchmarking or Prediction?	Time Serious Data?	Very Big Data?	Complex System?	Selected Algorithm?	Append Clusters Labels to Data? or Normalize Data?	Training, Validation and Tuning	Evaluation	
Not asked	—	√	√	—	—	—	prediction	√	√	√	GPR	normalized	Split data to train and test with different samples steps	Used test data with different samples steps	[33]
Too many features	Selected features using *p*-value	×	√	—	—	replaced	Benchmarking	×	√	Not asked	k-mean	normalized	Applied similarity measure on one package data	Compare results with EnergyStar software	[37]

Table 2. The details of the reviewed papers and the comments that result from comparison with the proposed ML pipeline.

Model Target	Data Source	Data Size	Model Features	Selected Algorithm?	Best Evaluation Results	Comments and Expected Improvements
Predict Hourly Energy Consumption [21]	"Two different data sets provided from two different buildings: The first set is the benchmark PROBEN 1, and comes from the first energy prediction contest, the Great Building Energy Predictor Shootout I, organized by ASHRAE (data set A) & The second data set derives from an office building located in Athens, Greece (data set B)"	data set A: a total of 4208 time steps, data set B: a total of 8280 time steps	data set A: "temperature, solar radiation, humidity ratio and wind speed" data set B: "ambient temperature, humidity, daily, weekly and yearly cycles the hour of day, day of week and day of year"	ANN	data set A: RMS is 15.25, MAPE is 1.50, CV is 2.44 and MBE is 0.37 data set B: RMS is 1.13, MAPE is 2.64, CV is 2.95 and MBE is –0.03	There are some wrong prediction peaks due to ignoring the effect of the outliers in preprocessing data step, and the evaluation would be better if carried out on the same model structures with different data sets but with the same input features to increase reliability and robustness.

Table 2. *Cont.*

Model Target	Data Source	Data Size	Model Features	Selected Algorithm?	Best Evaluation Results	Comments and Expected Improvements
Predict Hourly Heating Energy [13]	"A model house designed in Denizli which is located in Central Aegean Region of Turkey"	A total of 35,070 time steps	"Month, day of the month, hour of the day, and energy consumption values at certain hours"	ANN	RMSE is 1.2125, R2 is 0.9880 and MAPE is 0.2081	The author did not mention any preprocessing steps, just normalization because the used data was calculated, which did not have the same worth of actual data, and the trained model could be more robust if the cross-validation technique was used in training and different time steps during the evaluation,
Predict Hourly Energy Consumption [57]	"CIESOL bioclimatic building, located in the southeast of Spain"	A total of 700,000 time steps	"The type and hour of the day, weather variables (outdoor temperature, outdoor humidity, solar radiation, wind velocity and wind direction) and the state of the actuators from the solar cooling installation"	ANN	Mean error is 11.48%	Although the authors mentioned the outliers and noise in the data, they did not apply any type of analysis to solve these two problems in the data. In addition, the missing values in the data are kept as is and the authors depended on a manual method in splitting data to skip missing values, which means the splitting blocks are imbalanced. Therefore, the efforts made in the training and evaluation steps to create a robust model were useless because the preprocessing steps are not well performed, so the results from the model have a relatively high mean error.
Predict Hourly electricity consumption [16]	"The Great Building Energy Predictor Shootout I, organized by ASHRAE in 1990s (data set A) Data from a library building located in Hangzhou, East China (data set B)"	data set A: a total of 4208 time steps, data set B: a total of 2472 time steps	Data A: "outdoor dry bulb temperature, solar radiation, humidity ratio and wind speed" Data B: "daily temperature and occupancy"	ANN	data set A: CV is 0.0254 and MAPE is 0.0162 data set B: CV is 0.0758 and MAPE is 0.058	The authors did not mention anything about the missing values in the data. The automatic tuning gives high prediction results, but it needs to integrate with a cross-validation technique to ensure the robustness and reliability for model.

Table 2. *Cont.*

Model Target	Data Source	Data Size	Model Features	Selected Algorithm?	Best Evaluation Results	Comments and Expected Improvements
Benchmark and predict (hourly and weekly) Energy consumption [36]	"10 institutional buildings in Singapore"	a total of 122 days for each building	"Hourly and weekly energy consumption"	k-shape and SVM	Respective MAPE values are 15.36, 9.46, 1.033 1.23, 2.37, 3.66, 0.57, 54.11, 3.63, 4.46 for the ten buildings	The authors mentioned the noise in data but did not mention solving it. In addition, the outliers and missing values did not mention solve in the technique. The data size is relatively high to be used in SVM algorithms (the authors did not take into consideration the data size when selecting the algorithm), and the authors extracted features to decrease complexity and effort. Thus, it is recommended to used parallel SVM to reduce time.
Predict Daily Energy Performance [33]	"Real weather data in the Chicago area"	a total of 8736 time steps	"Weather and occupancy levels, and the most commonly used is outdoor dry-bulb air temperature"	GPR	SSE is from 2.7e5 to 3.6e6 and total energy savings prediction error is from 31 to 41.23	The data size is relatively large, and the authors did not mention time consumption in training, as it may be too large. Therefore, it will be better to use the Gaussian model to remove noise only and complete prediction by ANN or use ANN directly for all problems.
Benchmark annual Energy Performance [37]	"commercial building (CBECS database)"	5215 samples	"Area, percent heated, percent cooled, wall materials, roof materials, window materials, window percent, shape, number of floors, construction year, weekly operation hours, occupants, variable air volume, heating unit, cooling unit, economizer, refrigerators, number of servers, office equipment, heating and cooling degree day"	k-mean	Ratio between actual energy index to centroid for cluster in range from 0.96 to 2.1 for each cluster	The data contain outliers, but the authors did not mention this or the technique to solve this. In addition, the imputation for missing values is not declared well. The evaluation step is carried out using a comparison with EnergyStar without declaration of any approach to overcome the local minimum solution of the k-mean algorithm.

4. Implement the Pipeline on CBECS Data

In this section, the pipeline is used as a reference to make commercial building energy predictions by using CBECS data. The data are collected by the US Energy Information Administration (EIA). Since 1979, the EIA has performed the CBECS regularly, as mandated by Congress. For commercial buildings, the EIA gathers data in two parts: (1) building characteristics or features are gathered through an in-person or online survey of building owners and managers, and (2) energy use data are gathered from power suppliers.

After collecting the data, the preprocessing and visualization step follows. The targets during data visualization are to (1) check missing values, (2) check outlier values, and (3) understand the nature of each feature distribution (normal distribution or skewed distribution). The visualization helps to choose the best method suitable for filling missing values and replacing outlier values. The difference between mean and median reflects the influence of outliers on data distribution, as seen by the calculation procedure of mean and median values for each characteristic in Table 3. In addition, the visualization of missing values of each feature is very important to decide which feature is suitable to be taken in the training of ML because features with a high percentage of missing values cannot be taken.

Table 3. Selected features' characteristics.

Selected Features	Values and Ranges Format	Analysis before Changes	Notes and Changes	Analysis after Changes
Square footage (SQFT)	1001–1,500,000	Mean = 124,473.50 Median = 20,750.00 Std = 258,613.18 Outliers = 12.31% Missing = 0.0%	No changes	Mean = 124,473.50 Median = 20,750.00 Std = 258,613.18 Outliers = 12.31% Missing = 0.0%
Number of floors (NFLOOR)	1–14 994 = 15 to 25 995 = More than 25	Mean = 30.16 Median = 2.00 Std = 163.61 Outliers = 9.73% Missing = 0.0%	Change (994 = 15 to 25) to ('20' = 15 to 25) as mean value to this range and change (995 = More than 25) to (30 = More than 25) [17]	Mean = 3.01 Median = 2.00 Std = 4.31 Outliers = 9.73% Missing = 0.0%
Year of construction (YRCON)	995 = Before 1946 1946–2012	Mean = 1861.10 Median = 1981.00 Std = 325.77 Outliers = 12.37% Missing = 0.0%	Change (995 = Before 1946) to (1932 = Before 1946)	Mean = 1976.97 Median = 1981.00 Std = 23.34 Outliers = 0.00% Missing = 0.0%
Total hours open per week (WKHRS)	0–168	Mean = 78.02 Median = 60.00 Std = 51.37 Outliers = 0.00% Missing = 0.0%	No changes	Mean = 78.02 Median = 60.00 Std = 51.37 Outliers = 0.00% Missing = 0.0%
Number of employees (NWKER)	0–6500	Mean = 178.78 Median = 15.00 Std = 565.94 Outliers = 15.97% Missing = 0.0%	No changes	Mean = 178.78 Median = 15.00 Std = 565.94 Outliers = 15.97% Missing = 0.0%
Percent heated (HEATP)	0–100 Missing = Not applicable	Mean = 88.52 Median = 100.00 Std = 24.24 Outliers = 19.94% Missing = 7.75%	Fill missing values with 0 because (not applicable mean zero percentage)	Mean = 81.49 Median = 100.00 Std = 33.38 Outliers = 15.73% Missing = 0.0%

Table 3. *Cont.*

Selected Features	Values and Ranges Format	Analysis before Changes	Notes and Changes	Analysis after Changes
Percent cooled (COOLP)	1–100 Missing = Not applicable	Mean = 79.81 Median = 100.00 Std = 30.13 Outliers = 8.29% Missing = 10.18%	Fill missing values with 0 because (not applicable mean zero percentage)	Mean = 71.68 Median = 95.00 Std = 37.39 Outliers = 0.00% Missing = 0.0%
Number of computers (PCTERMN)	0–4195 Missing = Not applicable	Mean = 168.93 Median = 10.00 Std = 530.48 Outliers = 16.21% Missing = 3.56%	Fill missing values with 0 because (not applicable mean zero)	Mean = 162.92 Median = 9.00 Std = 521.90 Outliers = 16.1% Missing = 0.0%
Percent lit when open (LTOHRP)	0–100 Missing = Not applicable	Mean = 82.12 Median = 95.00 Std = 25.01 Outliers = 8.04% Missing = 4.4%	Fill missing values with 0 because (not applicable mean zero)	Mean = 78.50 Median = 90.00 Std = 29.70 Outliers = 8.66% Missing = 0.0%
Annual electricity consumption (thous Btu) (ELBTU)	Output Feature	Mean = 9,283,680.98 Median = 822,346.50 Std = 32,174,631.57 Outliers = 14.67% Missing = 2.47%	No changes	Mean = 9,283,680.98 Median = 822,346.50 Std = 32,174,631.57 Outliers = 14.67% Missing = 2.47%

The first question in the pipeline comes after the preprocessing and visualization step (i.e., do the data have a high number of features?). The data contain a large number of features, so the feature selection step must be implemented to select the most significant features. The features selection step depends on the paper [17], which depended on calculating linear correlations between studied features and selecting the most appropriate ones (i.e., have a low level of missing values and high correlation with output features). The selected features are ('Square footage', 'Number of floors', 'Year of construction', 'Total hours open per week', 'Number of employees', 'Percent heated', 'Percent cooled', 'Number of computers', 'Percent lit when open', 'Annual electricity consumption (thous Btu)').

By moving through the pipeline, the answer to the next question (i.e., are the data noisy?) leads to the final part in the preprocessing steps (i.e., solving the problems of missing values and outliers). These problems are primarily declared through the visualization step for features. There are missing values and outliers in some features, so some changes are made (shown in Table 3) as a first action to reduce these effects during ML model training. After making these changes, the missing values are eliminated in all features except the 'Annual electricity consumption (thous Btu)' feature, which still has a small percentage of missing values that can be replaced by the median value of the feature. In addition, the percentage of outliers' values decreased significantly but was not eliminated.

The remaining outliers' values can be decreased by combining two features or more in one feature to reduce the effect of very high and very low values in each feature on the model. The final features are ('Total hours open per week', 'Building age', 'Building area per employee', 'Building area per PC', 'Building area per employee', 'Number of floors', 'Percent heated', 'Percent cooled', 'Percent lit when open', and 'Electricity use per area'). The new features' analysis shows that the new features (Table 4) have low outlier percentages compared with the original features in Table 3. Some of the new features have left-skewed distribution and a high percentage of outliers. Therefore, the log-scale transformation can be used to reduce the effect of outliers' values on the ML model.

Table 4. The final features of the ML model.

Selected Features	Analysis before Changes	Notes and Changes	Analysis after Changes
Total hours open per week	Mean = 78.02 Median = 60.00 Std = 51.37 Outliers = 0.00%	No changes	Mean = 78.02 Median = 60.00 Std = 51.37 Outliers = 0.00%
Building age	Mean = 35.03 Median = 31.00 Std = 23.34 Outliers = 0.00%		Mean = 35.03 Median = 31.00 Std = 23.34 Outliers = 0.00%
Building area per employee	Mean = 596,141.43 Median = 1176.48 Std = 2,362,836.96 Outliers = 13.33%		Mean = 7.62 Median = 7.07 Std = 2.37 Outliers = 6.92%
Building area per PC	Mean = 1,530,342.86 Median = 2000.00 Std = 3,595,546.34 Outliers = 18.51%	Convert to log scale	Mean = 8.72 Median = 7.60 Std = 3.35 Outliers = 15.28%
Number of floors	Mean = 3.01 Median = 2.00 Std = 4.31 Outliers = 9.73%		Mean = 0.65 Median = 0.69 Std = 0.82 Outliers = 2.80%
Percent heated	Mean = 81.49 Median = 100.00 Std = 33.38 Outliers = 15.73%		Mean = 81.49 Median = 100.00 Std = 33.38 Outliers = 15.73%
Percent cooled	Mean = 71.68 Median = 95.00 Std = 37.39 Outliers = 0.00%	No changes	Mean = 71.68 Median = 95.00 Std = 37.39 Outliers = 0.00%
Percent lit when open	Mean = 78.50 Median = 90.00 Std = 29.70 Outliers = 8.66%		Mean = 78.50 Median = 90.00 Std = 29.70 Outliers = 8.66%
Electricity use (thous Btu) per area	Mean = 64.29 Median = 40.96 Std = 82.96 Outliers = 7.16%	Convert to log scale	Mean = 3.58 Median = 3.71 Std = 1.24 Outliers = 2.83%

Finally, the remaining outliers' values are deleted and the final size of used data after all preprocessing steps is 4371 samples. Because of the large data size, the selected algorithm step is the ANN algorithm. Thus, the data will be normalized by each maximum value in each feature [17].

In the pipeline, after selecting the appropriate ML model, two steps must be performed: (1) using k-fold cross-validation in model training, validation, and hyper-parameters tuning steps; and (2) a model evaluation step by using unseen data. The ANN model has hyper-parameters such as (1) the learning rate, (2) the number of dense layers, and (3) the number of nodes per layer. Because of difficulties in tuning hyper-parameters, different combinations of hyper-parameters are used and evaluated by using adjusted R-squared values during the evaluation step on test data (Table 5). Some other hyper-parameters are fixed for all models such as (1) mean square error (MSE) as a loss function, (2) stochastic gradient descent (SGD) as an optimizer and (3) k = 5 for the cross-validation technique.

Table 5. Results of different ANN architectures (Model 7 achieves best results).

ANN Models	Hyper-Parameters			Test Results
	Learning Rate	**Dense Layers Number**	**Nodes Number**	**Adjusted R2**
Model 1	4.54×10^{-3}	1	143	0.63
Model 2	2.31×10^{-3}	1	512	0.64
Model 3	2.18×10^{-5}	5	434	0.45
Model 4	4.98×10^{-3}	7	119	0.76
Model 5	3.28×10^{-3}	21	124	0.85
Model 6	9.98×10^{-5}	25	227	0.9
Model 7	$\mathbf{9.41 \times 10^{-5}}$	**25**	**263**	**0.91**
Model 8	40.47×10^{-5}	30	180	0.897
Model 9	29.79×10^{-5}	30	74	0.894

The results of different ANN architectures are shown in Table 5, where model 7 achieves the best results on test data and can be deployed in real life as discussed in the ML pipeline. The values of hyper-parameters for different models declare that: (1) deeper ANN can obtain higher prediction results, but too many layers reduce results; (2) the small value for learning rate makes model 3 fall into local minimum; (3) the high value for learning rate makes the models 1, 2, 4, and 5 fluctuate around minimum loss during training; and (4) the change of nodes per layer does not have the same significant effect as changing the number of layers.

5. Conclusions

This paper overcomes the problem of losing experience in the ML concepts and applications by (1) providing an explanation for the building energy applications of ML over the world to increase knowledge of applications and (2) a clear explanation for advantages and drawbacks of each reviewed ML algorithm and how to implement each one to achieve the highest performance, and by (3) proposing a generic ML pipeline for the energy and building field with recommended preprocessing steps. In addition, the steps of implementing ML algorithms are very clear: (1) select and justify the appropriate ML approach for a given problem such as benchmarking or prediction; (2) build, train, evaluate, deploy, and fine-tune a machine learning model; (3) apply the steps of the ML pipeline to solve a specific problem; (4) describe some of the best practices for designing scalable, cost-optimized, and reliable models; and (5) identify the steps needed to apply machine learning in real life.

As its first contribution, this paper proposes in Figure 1 ML building energy applications for ANN, SVM, GPR or GMM, and Clustering algorithms, which include (1) energy assessment studies, (2) prediction for loads and energy consumption, (3) classification of energy consumption in buildings, (4) modeling solar radiation and solar steam generators, (5) modeling and forecasting loads for air conditioning systems, (6) simulating and controlling for energy consumption systems, (7) fault detection and diagnosis, and (8) energy saving, verification, and retrofit studies.

The second contribution of this paper is the general ML pipeline (Figure 2) to be used in the energy and building domain, which summarizes the requirements for each ML algorithm used depending on reviewed papers and how to overcome the drawbacks of each one. The pipeline is as follows: (1) identifying a real-life problem such as building prediction or benchmarking, (2) the real-life problem is transformed into an ML problem during the problem formulation step, and (3) the data about the problem must be collected to cover different cases of the problem and integrated if collected from different resources.

(4) The visualization step for the collected data helps to study the nature of problem by data analysis, answering some questions concerning data size, data features, correlation between features, density distribution of data, mean, median, and mode of data and percentage of noise, missing values, and outliers. (5) The preprocessing step depends on the data analysis step and is where the best technique of preparing data is selected and outliers and missing values are removed. (6) The selection of ML algorithms depends on the data analysis step to answer the questions that identify the suitable algorithm for the problem. (7) The training, validation, and hyper-parameters' tuning process must be carried out depending on k-fold cross validation to cover all data points without falling into the local minimum problem or overfitting and underfitting problems. (8) The evaluation step is the key to knowing the overall performance of the trained model. Finally, depending on the model evaluation, (9) we choose between implementing the model in real life and monitoring its performance by a verification step, or rearranging the features and increasing the data sample.

The proposed pipeline lays out the main steps of evaluating any research in the energy and buildings field to identify the value of new research. This approach will reconsider all previous papers in the field to repeat the previous work with a declaration for implementation steps by using the ML pipeline to improve performance. In addition, it will reduce the time required for any new ML user who does not have enough experience in ML applications to enhance their work in a well-arranged pipeline. The contributions of this paper are approved through implementing the ML pipeline on a real case study (i.e., CBECS data), which helps in creating a robust ANN model for a real-life problem and evaluating the performance for each hyper-parameter to achieve the best results. During implementation, the pipeline represents an effective reference in handling one of the real-life problems (i.e., energy prediction for commercial buildings) and converting it to an ML model scientifically. The implementation finds that many steps are heavily repeated throughout the solving of ML problems. Putting these steps in one generic pipeline to deploy the right algorithms seamlessly, reduce the complexity of quickly transferring ML models into real life, and manage ML models easier increases the performance and organization of creating a scientific model for real-life problems in a sufficient way.

This paper may be the basis for benchmarking in the field of energy and buildings as well as for prediction software that need the user to select one application from the applications list and upload just two files containing input data and output data. Then, the software performs some statistical analysis to collect information about data such as data type (i.e., categorical or numerical and time serious or not), size, most significant input features, degree of noise, and mean, median, percentage of missing values, and outliers. From the selected application and all calculated information, the software will detect the best techniques required in each situation, choose algorithms, perform automatic training, tuning, and evaluation, and finally give the user the final model with its specifications file (i.e., model type and its inputs and outputs) to deploy it in real life. This future dream will decrease the effort and time required by engineering to solve such problems. The software may also have the energy certificates and regulations to conduct energy retrofit studies for users.

The dream of creating compatible software may have extended to other fields to create other versions; each one related to a specific field, but all are based on this ML pipeline. These types of software are very helpful nowadays to control our life more and more by managing real-life problems in some models in a fast, accurate, and scientific way.

Author Contributions: Conceptualization, M.A.B.A.; Formal analysis, M.A.B.A.; Investigation, M.A.B.A.; Methodology, M.A.B.A.; Software, M.A.B.A.; Supervision, M.H.; Validation, M.A.B.A.; Visualization, M.A.B.A.; Writing—original draft, M.A.B.A. Both authors have read and agreed to the published version of the manuscript.

Funding: The results are done by integrating graphical processing unit (NVIDA Tesla T4 GPU) with supercomputer specifications (PowerEdge R7525 & 2 x 64-Core 2.45 GHz AMD EPYC 7763 64-Core Processor & 512 GB memory) of Norwegian University of Science and Technology (NTNU).

Institutional Review Board Statement: Not applicable.

Informed Consent Statement: Not applicable.

Data Availability Statement: https://www.eia.gov/consumption/commercial/.

Conflicts of Interest: The authors declare no conflict of interest.

References

1. Zhao, H. Artificial Intelligence Models for Large Scale Buildings Energy Consumption Analysis. Ph.D. Thesis, Ecole Centrale Paris, Gif-sur-Yvette, France, 2014.
2. Tabrizchi, H.; Javidi, M.M.; Amirzadeh, V. Estimates of residential building energy consumption using a multi-verse optimizer-based support vector machine with k-fold cross-validation. *Evol. Syst.* **2019**. [CrossRef]
3. Cai, H.; Shen, S.; Lin, Q.; Li, X.; Xiao, H. Predicting the energy consumption of residential buildings for regional electricity supply-side and demand-side management. *IEEE Access* **2019**, *7*, 30386–30397. [CrossRef]
4. Seyedzadeh, S.; Rahimian, F.P.; Oliver, S.; Rodriguez, S.; Glesk, I. Machine learning modelling for predicting non-domestic buildings energy performance: A model to support deep energy retrofit decision-making. *Appl. Energy* **2020**, *279*, 115908. [CrossRef]
5. Somu, N.; Raman, G.R.M.; Ramamritham, K. A deep learning framework for building energy consumption forecast. *Renew. Sustain. Energy Rev.* **2021**, *137*, 110591. [CrossRef]
6. Fayaz, M.; Kim, D. A Prediction Methodology of Energy Consumption Based on Deep Extreme Learning Machine and Comparative Analysis in Residential Buildings. *Electronics* **2018**, *7*, 222. [CrossRef]
7. Liu, Z.; Wu, D.; Liu, Y.; Han, Z.; Lun, L.; Gao, J.; Cao, G. Accuracy analyses and model comparison of machine learning adopted in building energy consumption prediction. *Energy Explor. Exploit.* **2019**, *37*, 1426–1451. [CrossRef]
8. Wang, L.; El-Gohary, N.M. *Machine-Learning-Based Model for Supporting Energy Performance Benchmarking for Office Buildings*; Springer: Cham, Switzerland, 2018. [CrossRef]
9. Seyedzadeh, S.; Rahimian, F.P.; Glesk, I.; Roper, M. Machine learning for estimation of building energy consumption and performance: A review. *Vis. Eng.* **2018**, *6*, 5. [CrossRef]
10. Shalev-Shwartz, S.; Ben-David, S. Understanding Machine Learning_From Theory to Algorithm. 2014. Available online: https://www.cs.huji.ac.il/~{}shais/UnderstandingMachineLearning/understanding-machine-learning-theory-algorithms.pdf (accessed on 30 August 2021).
11. Chao, W.-L. Machine Learning Tutorial. 2011. Available online: https://www.semanticscholar.org/paper/Machine-Learning-Tutorial-Chao/e74d94c407b599947f9e6262540b402c568674f6 (accessed on 30 August 2021).
12. Kirsch, J.H.A.D. IBM Machine Learning for Dummies. 2018. Available online: https://www.ibm.com/downloads/cas/GB8ZMQZ3 (accessed on 30 August 2021).
13. Dombaycı, Ö.A. The prediction of heating energy consumption in a model house by using artificial neural networks in Denizli–Turkey. *Adv. Eng. Softw.* **2010**, *41*, 141–147. [CrossRef]
14. Antanasijević, D.; Pocajt, V.; Ristić, M.; Perić-Grujić, A. Modeling of energy consumption and related GHG (greenhouse gas) intensity and emissions in Europe using general regression neural networks. *Energy* **2015**, *84*, 816–824. [CrossRef]
15. Platon, R.; Dehkordi, V.R.; Martel, J. Hourly prediction of a building's electricity consumption using case-based reasoning, artificial neural networks and principal component analysis. *Energy Build.* **2015**, *92*, 10–18. [CrossRef]
16. Li, K.; Hu, C.; Liu, G.; Xue, W. Building's electricity consumption prediction using optimized artificial neural networks and principal component analysis. *Energy Build.* **2015**, *108*, 106–113. [CrossRef]
17. Yalcintas, M.; Aytun Ozturk, U. An energy benchmarking model based on artificial neural network method utilizing US Commercial Buildings Energy Consumption Survey (CBECS) database. *Int. J. Energy Res.* **2007**, *31*, 412–421. [CrossRef]
18. Edwards, R.E.; New, J.; Parker, L.E. Predicting future hourly residential electrical consumption: A machine learning case study. *Energy Build.* **2012**, *49*, 591–603. [CrossRef]
19. Kialashaki, A.; Reisel, J.R. Modeling of the energy demand of the residential sector in the United States using regression models and artificial neural networks. *Appl. Energy* **2013**, *108*, 271–280. [CrossRef]
20. Olofsson, T.; Andersson, S. Long-term energy demand predictions based on short-term measured data. *Energy Build.* **2001**, *33*, 85–91. [CrossRef]
21. Karatasou, S.; Santamouris, M.; Geros, V. Modeling and predicting building's energy use with artificial neural networks: Methods and results. *Energy Build.* **2006**, *38*, 949–958. [CrossRef]
22. Du, Z.; Fan, B.; Jin, X.; Chi, J. Fault detection and diagnosis for buildings and HVAC systems using combined neural networks and subtractive clustering analysis. *Build. Environ.* **2014**, *73*, 1–11. [CrossRef]
23. Huang, H.; Chen, L.; Hu, E. A neural network-based multi-zone modelling approach for predictive control system design in commercial buildings. *Energy Build.* **2015**, *97*, 86–97. [CrossRef]
24. Pérez-Ortiz, J.A.; Gers, F.A.; Eck, D.; Schmidhuber, J. Kalman filters improve LSTM network performance in problems unsolvable by traditional recurrent nets. *Neural Netw.* **2003**, *16*, 241–250. [CrossRef]
25. González, P.A.; Zamarreño, J.M. Prediction of hourly energy consumption in buildings based on a feedback artificial neural network. *Energy Build.* **2005**, *37*, 595–601. [CrossRef]

26. Aydinalp, M.; Ismet Ugursal, V.; Fung, A.S. Modeling of the space and domestic hot-water heating energy-consumption in the residential sector using neural networks. *Appl. Energy* **2004**, *79*, 159–178. [CrossRef]
27. Hou, Z.; Lian, Z. An application of support vector machines in cooling load prediction. In Proceedings of the 2009 International Workshop on Intelligent Systems and Applications, Wuhan, China, 23–24 May 2009.
28. Li, Q.; Meng, Q.; Cai, J.; Yoshino, H.; Mochida, A. Applying support vector machine to predict hourly cooling load in the building. *Appl. Energy* **2009**, *86*, 2249–2256. [CrossRef]
29. Li, X.; Lu, J.-H.; Ding, L.; Xu, G.; Li, J. Building Cooling Load Forecasting Model Based on LS-SVM. In Proceedings of the 2009 Asia-Pacific Conference on Information Processing, Shenzhen, China, 18–19 July 2009; pp. 55–58.
30. Jain, R.K.; Smith, K.M.; Culligan, P.J.; Taylor, J.E. Forecasting energy consumption of multi-family residential buildings using support vector regression: Investigating the impact of temporal and spatial monitoring granularity on performance accuracy. *Appl. Energy* **2014**, *123*, 168–178. [CrossRef]
31. Zhao, H.-X.; Magoulès, F. A review on the prediction of building energy consumption. *Renew. Sustain. Energy Rev.* **2012**, *16*, 3586–3592. [CrossRef]
32. Zhao, H.X.; Magoulès, F. Parallel Support Vector Machines Applied to the Prediction of Multiple Buildings Energy Consumption. *Algorithms Comput. Technol.* **2009**, *4*, 231–249. [CrossRef]
33. Heo, Y.; Zavala, V.M. Gaussian process modeling for measurement and verification of building energy savings. *Energy Build.* **2012**, *53*, 7–18. [CrossRef]
34. Burkhart, M.C.; Heo, Y.; Zavala, V.M. Measurement and verification of building systems under uncertain data: A Gaussian process modeling approach. *Energy Build.* **2014**, *75*, 189–198. [CrossRef]
35. Heo, Y.; Choudhary, R.; Augenbroe, G.A. Calibration of building energy models for retrofit analysis under uncertainty. *Energy Build.* **2012**, *47*, 550–560. [CrossRef]
36. Yang, J.; Ning, C.; Deb, C.; Zhang, F.; Cheong, D.; Lee, S.E.; Tham, K.W. k-Shape clustering algorithm for building energy usage patterns analysis and forecasting model accuracy improvement. *Energy Build.* **2017**, *146*, 27–37. [CrossRef]
37. Gao, X.; Malkawi, A. A new methodology for building energy performance benchmarking: An approach based on intelligent clustering algorithm. *Energy Build.* **2014**, *84*, 607–616. [CrossRef]
38. Lara, R.A.; Pernigotto, G.; Cappelletti, F.; Romagnoni, P.; Gasparella, A. Energy audit of schools by means of cluster analysis. *Energy Build.* **2015**, *95*, 160–171. [CrossRef]
39. Santamouris, M.; Mihalakakou, G.; Patargias, P.; Gaitani, N.; Sfakianaki, K.; Papaglastra, M.; Zerefos, S. Using intelligent clustering techniques to classify the energy performance of school buildings. *Energy Build.* **2007**, *39*, 45–51. [CrossRef]
40. Gaitani, N.; Lehmann, C.; Santamouris, M.; Mihalakakou, G.; Patargias, P. Using principal component and cluster analysis in the heating evaluation of the school building sector. *Appl. Energy* **2010**, *87*, 2079–2086. [CrossRef]
41. Kalogirou, S.A. Applications of artificial neural networks in energy systems a review. *Energy Convers. Manag.* **1998**, *40*, 1073–1087. [CrossRef]
42. Ascione, F.; Bianco, N.; De Stasio, C.; Mauro, G.M.; Vanoli, G.P. Artificial neural networks to predict energy performance and retrofit scenarios for any member of a building category: A novel approach. *Energy* **2017**, *118*, 999–1017. [CrossRef]
43. Beccali, M.; Ciulla, G.; Brano, V.L.; Galatioto, A.; Bonomolo, M. Artificial neural network decision support tool for assessment of the energy performance and the refurbishment actions for the non-residential building stock in Southern Italy. *Energy* **2017**, *137*, 1201–1218. [CrossRef]
44. Paudel, S.; Elmtiri, M.; Kling, W.L.; Le Corre, O.; Lacarrière, B. Pseudo dynamic transitional modeling of building heating energy demand using artificial neural network. *Energy Build.* **2014**, *70*, 81–93. [CrossRef]
45. Deb, C.; Eang, L.S.; Yang, J.; Santamouris, M. Forecasting diurnal cooling energy load for institutional buildings using Artificial Neural Networks. *Energy Build.* **2016**, *121*, 284–297. [CrossRef]
46. Benedetti, M.; Cesarotti, V.; Introna, V.; Serranti, J. Energy consumption control automation using Artificial Neural Networks and adaptive algorithms: Proposal of a new methodology and case study. *Appl. Energy* **2016**, *165*, 60–71. [CrossRef]
47. Ahn, J.; Cho, S.; Chung, D.H. Analysis of energy and control efficiencies of fuzzy logic and artificial neural network technologies in the heating energy supply system responding to the changes of user demands. *Appl. Energy* **2017**, *190*, 222–231. [CrossRef]
48. Kalogirou, S.; Lalot, S.; Florides, G.; Desmet, B. Development of a neural network-based fault diagnostic system for solar thermal applications. *Sol. Energy* **2008**, *82*, 164–172. [CrossRef]
49. Hong, S.-M.; Paterson, G.; Mumovic, D.; Steadman, P. Improved benchmarking comparability for energy consumption in schools. *Build. Res. Inf.* **2013**, *42*, 47–61. [CrossRef]
50. Buratti, C.; Barbanera, M.; Palladino, D. An original tool for checking energy performance and certification of buildings by means of Artificial Neural Networks. *Appl. Energy* **2014**, *120*, 125–132. [CrossRef]
51. Lai, F.; Magoulès, F.; Lherminier, F. Vapnik's learning theory applied to energy consumption forecasts in residential buildings. *Int. J. Comput. Math.* **2008**, *85*, 1563–1588. [CrossRef]
52. Li, X.; Bowers, C.P.; Schnier, T. Classification of Energy Consumption in Buildings with Outlier Detection. *IEEE Trans. Ind. Electron.* **2010**, *57*, 3639–3644. [CrossRef]
53. Oladipupo, T. Types of Machine Learning Algorithms. *New Adv. Mach. Learn.* **2010**, *3*, 19–48.
54. Wong, S.L.; Wan, K.K.W.; Lam, T.N.T. Artificial neural networks for energy analysis of office buildings with daylighting. *Appl. Energy* **2010**, *87*, 551–557. [CrossRef]

55. Smola, A.; Vishwanathan, S.V.N. *Introduction to Machine Learning*; Cambridge University: Cambridge, UK, 2008.
56. Deisenroth, M.P.; Faisal, A.A.; Ong, C.S. *Mathematics for Machine Learning*; Cambridge University Press: Cambridge, UK, 2020.
57. Mena, R.; Rodríguez, F.; Castilla, M.; Arahal, M.R. A prediction model based on neural networks for the energy consumption of a bioclimatic building. *Energy Build.* **2014**, *82*, 142–155. [CrossRef]

Forecasting Brazilian Ethanol Spot Prices Using LSTM

Gustavo Carvalho Santos [1,*,†], Flavio Barboza [2,*,†], Antônio Cláudio Paschoarelli Veiga [1,*,†], and Mateus Ferreira Silva [3,*,†]

[1] Electrical Engineering School, Federal University of Uberlândia, Uberlândia 38408-100, Brazil
[2] School of Business and Management, Federal University of Uberlândia, Uberlândia 38408-100, Brazil
[3] School of Accounting, Federal University of Uberlândia, Uberlândia 38408-100, Brazil
* Correspondence: gustavocavsantos@gmail.com (G.C.S.); flmbarboza@ufu.br (F.B.); acpveiga@ufu.br (A.C.P.V.); mateusferreira2@ufu.br (M.F.S.); Tel.: +55-34-3230-9472 (F.B.)
† These authors contributed equally to this work.

Abstract: Ethanol is one of the most used fuels in Brazil, which is the second-largest producer of this biofuel in the world. The uncertainty of price direction in the future increases the risk for agents operating in this market and can affect a dependent price chain, such as food and gasoline. This paper uses the architecture of recurrent neural networks—Long short-term memory (LSTM)—to predict Brazilian ethanol spot prices for three horizon-times (12, 6 and 3 months ahead). The proposed model is compared to three benchmark algorithms: Random Forest, SVM Linear and RBF. We evaluate statistical measures such as MSE (Mean Squared Error), MAPE (Mean Absolute Percentage Error), and accuracy to assess the algorithm robustness. Our findings suggest LSTM outperforms the other techniques in regression, considering both MSE and MAPE but SVM Linear is better to identify price trends. Concerning predictions per se, all errors increase during the pandemic period, reinforcing the challenge to identify patterns in crisis scenarios.

Keywords: price prediction; trend prediction; LSTM; SVM; Random Forest; MAPE; MSE; commodity price

Citation: Santos, G.C.; Barboza, F.; Veiga, A.C.P.; Silva, M.F. Forecasting Brazilian Ethanol Spot Prices Using LSTM. *Energies* **2021**, *14*, 7987. https://doi.org/10.3390/en14237987

Academic Editor: Ana-Belén Gil-González

Received: 27 October 2021
Accepted: 22 November 2021
Published: 30 November 2021

Publisher's Note: MDPI stays neutral with regard to jurisdictional claims in published maps and institutional affiliations.

1. Introduction

Ethanol has become an interesting alternative to fossil fuels in the world. In Brazil, this biofuel is widely used, and Brazilian production is the second largest in the world, behind only the United States. The importance of this biofuel in Brazil is because the country has succeeded in replacing oil with ethanol in 20% of the automotive fuel and thus 80% of Brazilian cars can carry several mixtures of gasoline and ethanol. This substitution took place due to the fact the country was severely affected by the 1973 oil crisis, in which the local government invested in an ambitious program "Proalcool" for the production of ethanol as a substitute for gas [1].

The main form used in the global industry for the production of ethanol is alcoholic fermentation through a microbiological process [2]. In Brazil, the main raw material for this process is sugarcane, where sugars are transformed into ethanol, energy, cell biomass, CO_2 and other secondary products by yeast cells [3].

In terms of economic impact, the international market of ethanol produced in Brazil is expressive and widespread. According to government data [4], the local producers export to more than 60 countries and imports to almost 20, and the country is among the top 3 exporters of this commodity [5]. Moreover, the Energy Information Administration (EIA, [6]) reports the importance of the Brazil-US relationship, as well as highlights political aspects that influence transactions in these countries.

Some studies revealed ethanol prices can affect many products, mainly food and gasoline [7]. Besides, it is not difficult to find researchers that find contrary outlines. For instance, David et al. [8] evaluates the price transmission via cointegration among ethanol and other commodities, especially coffee and Carpio [5] noticed the oil prices have affected ethanol prices when analysing more than 20 years. In addition, the volatility of spot prices encourages agents for hedging against market risk in the Brazilian and American markets [9].

The ethanol price is peculiar as long as it is not standardized for a global trade like oil, corn or coffee [5]. Although the spot price of ethanol in Brazil is given by the market, there are timid financial devices on the local stock exchange, which leads many producers and consumers to assume market risk. However, it has growth potential as renewable energy but also due to efficiency improvements in its production [10] and enormous current and future world necessity of energy [11]. Thus, as Brazil is one of the great markets [5,12], the behavior of local price is an element to be observed around the world.

Another issue involved here is the complexity of this task. Predicting price is one of the greatest challenges in the financial environment. There are obvious reasons (the future and dynamic nature) and methodologies employed for that. On the one hand, Statistical modelling needs to simplify the phenomena, when attempting to see linear structures. On the other hand, newer techniques arise from artificial intelligence development and demonstrated interesting outcomes [11,13–15] when outperformed statistical ones in many cases, notably for complex backgrounds.

Based on that, our study intends to use artificial neural networks with LSTM architecture to forecast the spot prices of the Brazilian market for this fuel based on sugar cane. To the best of our knowledge, this is the first study dedicated to forecast ethanol price. The results obtained in this paper demonstrate that it is possible to forecast ethanol prices in Brazilian sight with a degree of correctness of direction between 68 and 80% for the periods of 63, 125 and 252 working days (which is equivalent to 3, 6 and 12 months). Also, for validation of the algorithm, 3 models are compared to LSTM: Random Forest (RF), and Support Vector Machine (SVM)—with Linear and RBF kernels, since these techniques have shown satisfactory performance in the financial market contexts [13–15]. The model revealed interesting prediction power of the Brazilian biofuel price with a small error in periods of low volatility but poor performance occurred during the pandemic caused by COVID-19 due to the sharp drop of the commodity prices in this period.

This study contributes to the literature while it is the first study that examines machine learning models for forecasting Brazilian Ethanol, especially LSTM networks. The implemented algorithms can help practitioners to improve their performances, as well as enable the application of advanced strategies for hedging portfolios, as well as speculating ones.

The paper is organised in the following parts: In Section 2, we review related studies. Section 3 describes the proposed LSTM model and our methodology. The empirical results are presented in Section 4. The last part of the paper (Section 5) presents the concluding remarks and some recommendations.

2. Related Work

Several studies have been dedicated to applying statistical techniques in order to understand the behaviour of ethanol prices and establish dependency relationships in different markets. David et al. [9] used several tools such as Autoregressive Integrated Moving Average, Autoregressive Fractional Integrated Moving Average, Detrended Fluctuation Analyzes and Hurst and Lyapunov exponents to investigate the mechanism of ethanol prices in Brazil in the period from 2010 to 2015. According to the author, results demonstrate that the price of biofuel is antipersistent.

Bouri et al. [16] stated that the generalized autoregressive conditional heteroscedasticity (GARCH) models can incorporate structural breaks and improve the prediction of the volatility of the ethanol market in the United States. They also noted that the influence of good and bad news is properly assessed under such breaks.

It is also possible to find in the literature numerous papers that study the relationship between ethanol prices with other commodities: Carpio [5] relates the long-term and short-term effects of oil prices on ethanol, gasoline, and sugar price predictions. The author concludes that ethanol is sensitive to short- and long-term changes in the oil. David et al. [8] state that in general, ethanol has a lower predictability horizon than other commodities. Pokrivčák and Rajčaniová [17] also find a relationship in oil and ethanol prices. Bastianin et al. [18] suggest evidence that ethanol can be predicted by returns on corn.

However, studies involving artificial intelligence and the forecasting of ethanol prices are still scarce, despite the large number of works related to machine learning applied to commodity time series. In particular, Bildirici et al. [19] tested a hybrid model (GARCH + LSTM) to analyze the volatility of oil prices, including the effects of the COVID-19 pandemic. Their findings bring to light the contribution of LSTM, especially because of the complexity usually prevails in such data.

Dealing regression algorithms, Ding and Zhang [20] examined the effects of oil, copper, gold, corn, and cattle among them in terms of correlations. More specifically, the authors applied the cointegration method and found a link between oil and copper, and pieces of evidences connected with governments' impact in the other commodities markets.

Kulkarni and Haidar [21] developed an ANN model-based to forecast crude oil price trends. One interesting comment in this paper emphasizes the problematic use of econometric models can deliver "misleading outputs" due to robust assumptions required to them. In terms of results, they reached an impressive rate of 78% for predictions of oil price one day ahead.

In another use of neural networks applied to commodity prediction task, Alameer et al. [14] adopted an LSTM architecture to forecast coal price movements in Australia. Based on a large dataset (about 30 years) with monthly observations, the main findings are: LSTM is better than SVM and MLP when comparing RMSEs; and, there are correlations with other commodities, such as oil, natural gas, copper, gold, silver and iron. Still using LSTM, Liu et al. [22] combined the variational mode decomposition method and LSTM to construct a forecasting model for non-ferrous metals prices. They achieved remarkable performance close to 95% of correctly price trends for Zinc, Copper and Aluminum by working with the 30th last prices to predict the next day as inputs.

Other studies have brought relevant progress to the literature in this field. For example, Herrera et al. [11] compares neural networks and autoregressive integrated moving average (ARIMA) in forecasting Cattle and Wheat prices. Hu et al. [23] implemented a hybrid deep learning approach by integrating LSTM networks with the generalized autoregressive conditional heteroskedasticity (GARCH) model for copper price volatility prediction. Zhou et al. [24] uses a hybrid classification framework to forecast the price trend of bulk commodities over upcoming days, results show an f-score of up to 82%; Ouyang et al. [25] uses long- and short-term time series network for agricultural commodity futures prices prediction.

The papers cited demonstrate several techniques used for analyzing and forecasting commodities, in addition to studying the correlations of different assets with each other and their effects on the world and local economy. Thus, observing the papers developed on the topic of commodity price prediction using artificial intelligence, it is possible to verify a predominance of neural network algorithms, especially the implementation of the LSTM architecture [14,19,22,23].

3. Methodology

3.1. Data

The Center for Advanced Studies on Applied Economic (CEPEA) is an economic research department at Luiz de Queiroz School of Agriculture (ESALQ) from the University of São Paulo (USP) that gathers and provides data from economic, financial, social and environmental aspects of about 30 agribusiness supply chains [26]. The time series analyzed in this research holds daily prices of hydrous ethanol, collected from CEPEA/ESALQ/USP database, which covers the period from 25 January 2010 to 11 December 2020. This time interval includes all data available for ethanol prices up to the conclusion of this research.

We chronologically separated the data in the proportion of 80% for training the neural network and 20% to validate the model. Figure 1 illustrates the prices for the period specified above.

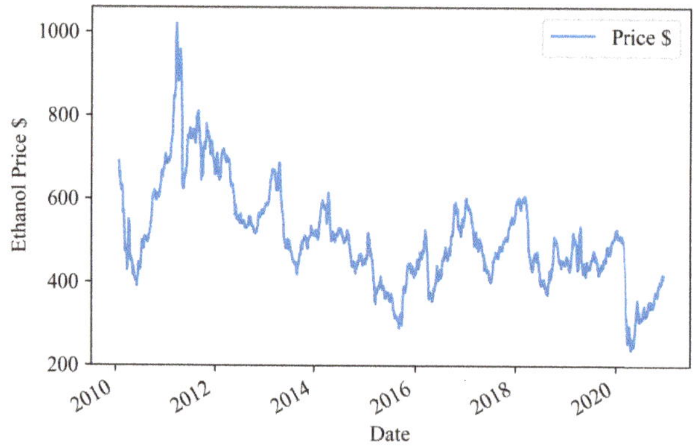

Figure 1. Brazilian Ethanol Spot Prices in U$ Dollars over the period between 2010 and 2020.

Data Pre-Processing

The inputs used in the proposed model vary according to the forecast horizon used. We use a rolling window containing the last $5\times$ days for each model. For example, let's assume we want to predict the price d days ahead. Then, we use data (Close Price of the ethanol in the day d) from d, $2d$, $3d$, $4d$, and $5d$ days before as inputs to the LSTM. In this paper, we apply 3 horizons in business days which are close to 3 and 6 months and 1 year of a calendar time. Table 1 shows the rolling windows used as inputs. These horizons are based on the required time for producing sugar cane, the main input of ethanol. One year covers the whole production [27] and shortest ones get partial perspective and can give the best point to hedge for anyone (buyers and sellers)[28].

Table 1. Steps used as features. C_t means Close Price at time t.

Forecast Horizon (Days)	Feature 1	Feature 2	Feature 3	Feature 4	Feature 5	Feature 6
63	C_t	C_{t-63}	C_{t-126}	C_{t-189}	C_{t-252}	C_{t-315}
126	C_t	C_{t-126}	C_{t-252}	C_{t-378}	C_{t-504}	C_{t-630}
252	C_t	C_{t-252}	C_{t-504}	C_{t-756}	C_{t-1008}	C_{t-1260}

In order to increase the efficiency of the predictors [23], the features were normalized through a StandardScaler algorithm provided by the Scikit-learn package. Basically, this scaler transforms the data into a normal distribution. It's important to note that the parameters of the distribution are given by the training sample and reused for transforming the test sample.

The price Y to be forecast on the Nth day is determined by looking at d days ahead to the current price C. Equation (1) shows this process:

$$Y_N = C_{N+d}.$$ (1)

3.2. LSTM Networks

A neural network is a data processing system that is based on the structure of brain neurons. Thus, it consists of a large number of simple processing and highly interconnected elements in an architecture [29]. There are several types of network architectures, this paper uses a model with dense and LSTM layers. This last type of architecture is widely used for learning sequences (time series, word processing and others) and is very sensitive when choosing hyperparameters [30]. According to Breuel [31] the performance of the LSTM slightly depends on the learning rate and the choice of non-linear recurrent activation functions (tanh and sigmoid) make the network perform better. Based on that, we chose sigmoid as the activation function.

Proposed by Hochreiter and Schmidhuber [32] as a solution for vanishing gradient problem and improved by Gers et al. [33] by introducing a forget gate into the cell, LSTM is a type of recurrent neural networks architecture. As in Yu et al. [34] based on Figure 2, the LSTM cell can be mathematically described as:

$$f_t = \sigma(W_f h_{t-1} + W_f x_t + b_f) \tag{2}$$

$$i_t = \sigma(W_i h_{t-1} + W_i x_t + b_i) \tag{3}$$

$$\tilde{c}_t = \sigma(W_c h_{t-1} + W_c x_t + b_c) \tag{4}$$

$$c_t = f_t \cdot c_{t-1} + i_t \cdot \tilde{c}_t \tag{5}$$

$$o_t = \sigma(W_o h_{t-1} + W_o x_t + b_o) \tag{6}$$

$$h_t = o_t \cdot tanh(c_t) \tag{7}$$

where f_t represents the forget gate, which allows the LSTM to reset its state [35]. When the f_t value is 1, it keeps that information, while the value 0 means that it deletes all that information. The input, the recurrent information, and the output of the cell at time t is portrayed by x_t, h_t and y_t respectively. The biases represented by b, i_t and o_t represent the input and output gates at time t. The cell state is symbolized by c_t and W_i, W_c, W_o and W_f are the weights. The operator '·' expresses the pointwise multiplication of two vectors.

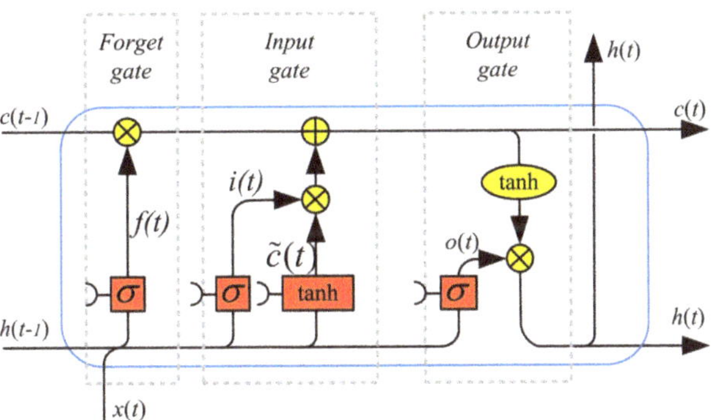

Figure 2. Architecture of LSTM with a forget gate. Reproduced from [34].

LSTM networks have a wide range of applications. It is possible to find in the literature several works in different areas that use this type of recurrent network to build machine learning models. Due to the ability to learn data sequences, numerous papers use this architecture for language processing and text classification [36–41], financial predictions [23,42–45], and other problems involving time series [19,46–50].

3.3. Proposed Model and Benchmarks

The model implemented in this work run in Python 3.8.6 (64-bit) with Jupyter Notebook. The hardware setup includes an Intel Core i5-4310u CPU 2.0GHz, 8GB RAM. The neural network is built using Tensorflow with Keras version 2.3.1 as interface.

The model's architecture includes four hidden layers with 64 units interspersed by a dropout of 20%, the first three are of the LSTM type and the last is dense. The purpose of the dropout layers is to randomly drop units from the neural network during training to avoid overfitting [51]. This architecture is also known as *vanilla LSTM* and has been applied in similar contexts [35,52].

Figure 3 illustrates an example of the implemented neural network architecture:

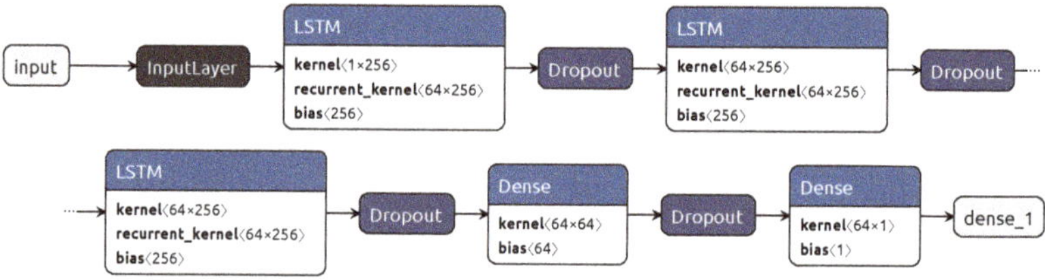

Figure 3. Neural network architecture.

In attempt to compare performances, we evaluate other 3 models: Random Forest (RF), Support Vector Machine (SVM) with two kernels: Linear (SVML) and Radial Basis Function (SVMR), which are considered as suggested techniques for this kind of problem [14,15]. Both are considered machine learning techniques [14,53]. RF is based on a collection of decision trees, in which classifies each instance by majority vote while SVM builds a hyperplane in attempt to optimize the division between classes.

We run all models in Python either and supported to scikit-learn libraries. In particular, we used parameters cost C = 1 and $\epsilon = 0.2$ to SVM. According to Carrasco et al. [54], the parameter C is responsible for the regularization, focusing in to avoid large coefficients and then contributing to lower misclassification rates, and epsilon is the width of the region (also called tube) centered in the hyperplane. This procedure tends to prevent overfitting. Other parameters remained as default settings. In the case of RF, all features keep as standard.

To check the error of the predictions made, the root mean squared error (RMSE) and the mean absolute percentage error (MAPE) were used. They can be defined as:

$$\text{RMSE} = \sqrt{\frac{1}{n} \sum (y_t - \hat{y}_t)^2} \tag{8}$$

$$\text{MAPE} = \frac{1}{n} \sum \frac{|y_t - \hat{y}_t|}{y_t} \tag{9}$$

where $\hat{y}_t$ is the forecast price and y_t the actual value, both in the time t, and n represents the number of forecast observations in the sample.

Another way to evaluate the predictions is by observing the ability of the models to adjust the change in price direction. We used the accuracy, precision and recall measures to do it. As in Wang et al. [55], these measures can be defined as:

$$\text{precision} = \frac{\text{correct predictions as } x}{\text{total predictions as } x} \tag{10}$$

$$\text{recall} = \frac{\text{correct predictions as } x}{\text{number of actual } x} \tag{11}$$

$$\text{accuracy} = \frac{\text{correct predictions}}{\text{total of predictions}} \tag{12}$$

where x can be the Upward or Downward trend.

4. Results and Discussions

First, we evaluate the model performance with the concern of detecting any bias in there. Second, we present the outputs and discuss them. Lastly, the visual analysis complements our investigation.

4.1. Learning Curves

Learning curves are a way to assess the ability of a deep learning model to generalize the realized information in the training phase. The curves of training and validation errors are plotted in Figures 4–6, representing short, medium and long-term horizons, respectively. In these cases, the curves can be observed through the number of epochs, which allows detecting possible overfitting of the model. If the curves of the training and validation errors decay together in a uniform trail, this issue can be discarded [56].

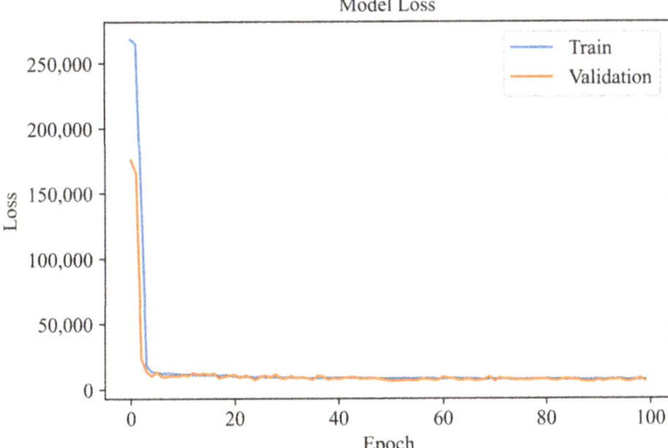

Figure 4. Loss for 63 days model.

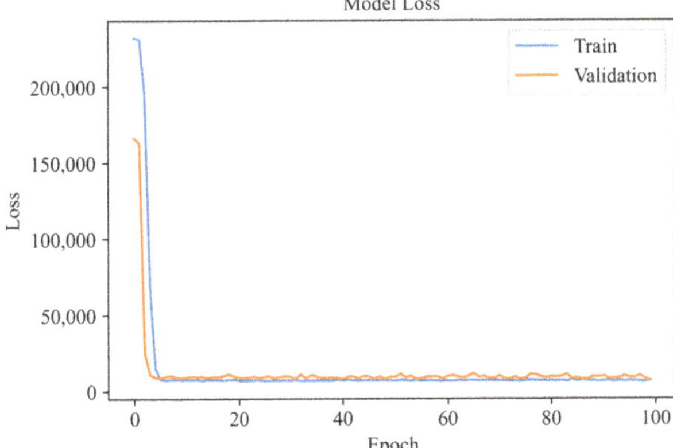

Figure 5. Loss for 126 days model.

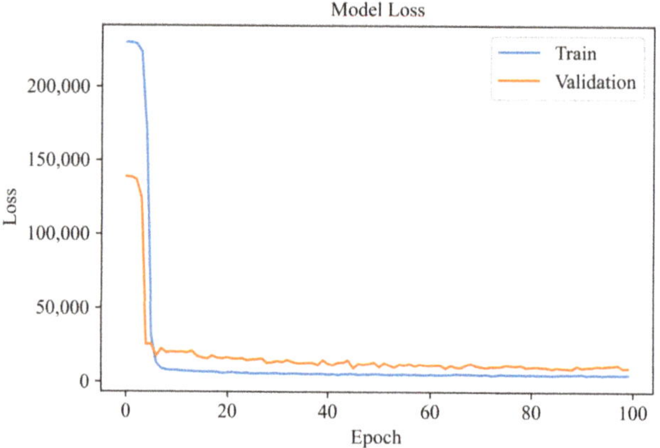

Figure 6. Loss for 252 days model.

Visually, the 63-day model obtained the best learning curves, and it is possible to observe that both error curves converged to a small value. In other words, this model is robust and sufficiently reliable. The same convergence can be viewed for the other models (126 and 252 days). However, it can be affirmed that the higher the forecast interval, the slower the error curves convergence.

The convergence of the learning curves of the three models presented coincide with their respective results in the validation: The 63-day model had the best convergence and also the smallest errors (MAPE and RMSE) in the predictions. On the other hand, the 252-days model had the lowest convergence and consequently the highest errors. These results can be verified in the next subsection.

4.2. Forecasting Results

We trained and validated the models of Long-Short Term Memory (LSTM), Random Forest (RF), Support Vector Machine with linear (SVML) and Radial Basis Function (SVMR) kernels, by using a chronological split of the full sample. Table 2 illustrates the error measures (MAPE and RMSE) obtained in the validation process for each forecast horizon.

Table 2. Error rates in each model for validation sample.

	63 Days Ahead		126 Days Ahead		252 Days Ahead	
Model	**MAPE**	**RMSE**	**MAPE**	**RMSE**	**MAPE**	**RMSE**
LSTM	17.23	78.53	19.91	83.38	26.15	98.69
RF	21.49	94.48	22.28	95.78	32.12	127.26
SVML	17.24	86.12	22.58	97.55	26.58	98.58
SVMR	20.65	92.62	23.32	98.00	33.72	120.22

It can be stated that the LSTM achieved lower errors (MAPE and RMSE) for all forecast horizons, except for the RMSE of the 252-day model. Similar to Alameer et al. [14], we can see the higher the average error increases the higher horizon-time. However, the results of our work show that the accuracy levels improve by extending the horizon of prediction (in the case of LSTM and SVML). These values show that the result is unwanted, but it is worth noting that such measures are commonly indicated for regression outlines. For example, an error rate of about 17% (on average, considering predictions in 63 days ahead) can be considered poorly contributory information and needs other measures to better understand the quality of the model. To do so, Table 3 provides details in terms of precision, recall and accuracy (hit ratios) to 63, 126, and 252 days, respectively, in this analysis.

Table 3. Hit ratios of predictions to 63-days horizon.

Model–Trend	Precision	Recall	Support	Accuracy
LSTM–Downtrend	0.59	0.88	173	72%
LSTM–Uptrend	0.90	0.63	291	
RF–Downtrend	0.80	0.65	173	81%
RF–Uptrend	0.81	0.90	291	
SVML–Downtrend	0.68	0.60	173	74%
SVML–Uptrend	0.78	0.86	291	
SVMR–Downtrend	0.74	0.84	173	83%
SVMR–Uptrend	0.90	0.82	291	

For 63-day forecasts, Table 3 shows that accuracy of 72% has been achieved, which can be considered a good result, and comparable to Zhou et al. [24]. On the one hand, the Upward movements obtained a better accuracy (90%), that is, of all predictions that pointed high, 9 out of 10 were correct. On the other hand, the recall of the Downward movements was higher (88%). In other words, out of every 100 occurrences of a fall in the asset value (actual values), the model presented 88 correct predictions. This result is can be considered good, but it is possible to improve. Hence, the optimized cutoff values can be found if any.

Table 4 shows that the LSTM has greater accuracy when sorting Uptrends for 126-horizon forecasts. This can be interpreted as bullish movements being clearer and less noisy signals. In addition, the total accuracy of the model was higher than the previous ones. On the other hand, for the 252-day forecasts, Table 5, although the total accuracy was 80%, the forecaster demonstrated to have low accuracy when classifying uptrend ethanol prices. This information casts doubt on the generality of LSTM to perform better in longer periods. Nevertheless, the only unbalanced sample was the latter (only 37 uptrend events against 200 downtrends). When looking specifically at this medium/long-term forecast, it is necessary to consider the analysis period, which includes moments not experienced by the training sample, like the pandemic caused by COVID-19. This effect probably justifies its poor performance in understanding the upwards and collaborates with the results of Bildirici et al. [19] when analyzing the impact of the pandemic on the prices of a commodity.

Table 4. Hit ratios of predictions to 126-days horizon.

Model–Trend	Precision	Recall	Support	Accuracy
LSTM–Downtrend	0.70	0.93	201	74%
LSTM–Uptrend	0.86	0.52	187	
RF–Downtrend	0.95	0.71	201	83%
RF–Uptrend	0.76	0.96	187	
SVML–Downtrend	0.84	0.78	201	81%
SVML–Uptrend	0.78	0.83	187	
SVMR–Downtrend	0.83	0.84	201	83%
SVMR–Uptrend	0.83	0.81	187	

Table 5. Hit ratios of predictions to 252-days horizon.

Model–Trend	Precision	Recall	Support	Accuracy
LSTM–Downtrend	0.88	0.88	200	80%
LSTM–Uptrend	0.35	0.35	37	
RF–Downtrend	0.87	0.40	200	44%
RF–Uptrend	0.17	0.68	37	
SVML–Downtrend	0.94	0.89	200	86%
SVML–Uptrend	0.53	0.70	37	
SVMR–Downtrend	0.97	0.50	200	57%
SVMR–Uptrend	0.25	0.92	37	

In general, these results show to be different from the work of Kulkarni and Haidar [21], in which the accuracy of the forecasts decays with the enlargement of the forecast horizon. However, it is important to highlight that in the cited article the forecast periods are only 1, 2 and 3 days. On the contrary, Bouri et al. [16] remind that the volatility (risk) perceived in periods is lower, which has greater meaning and more corresponding to this study.

Comparing LSTM outlines with benchmarks, its error rates are better, except for RMSE in the longer horizon (SVML is slightly better). Based on that, we can consider LSTM as the best predictor on a relative and regression basis. If we focus the analysis in the classification report, the findings are essentially unlike. LSTM presented just one single better indicator (recall in 63-day predictions, only bearing trends). Surprisingly, SVM achieved interesting performance in all horizons. For 63-day predictions, SVMR was the most accurate with the higher precision for uptrends. Curiously, the accuracy persisted to the mid-term horizon but decrease in the long-term while SVML was the best and reported two higher ratios (precision for uptrends and recall for downtrends). The RF precision and recall deserve to be emphasized. Except in the 252-day horizon, RF presented competitive ratios and more accurate than LSTM.

4.3. Visualising the Predictions

In the work of Bildirici et al. [19], it is possible to verify the impact of the COVID-19 pandemic on oil prices, in which the values suffered a great fall, affecting the entire international market. We can say the same for ethanol, as the big drop in prices also occurred. In addition, the author's analysis that prices will return to their highest values—potentially causing inflation—may also be valid for ethanol.

Another important point is the impact of the pandemic on forecast errors. As it is an adverse and unpredictable event, the error rates (MAPE and RMSE) were greater in this period compared to moments prior to the COVID-19 crisis.

Figures 7–9 illustrate the predictions of the algorithms versus the price verified in the analyzed period.

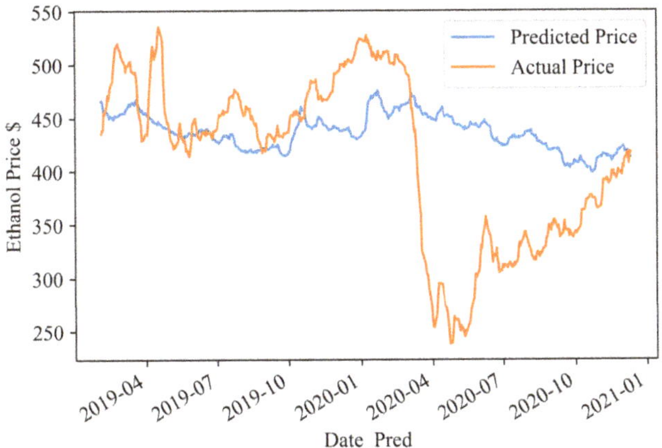

Figure 7. Predictions for 63 days ahead.

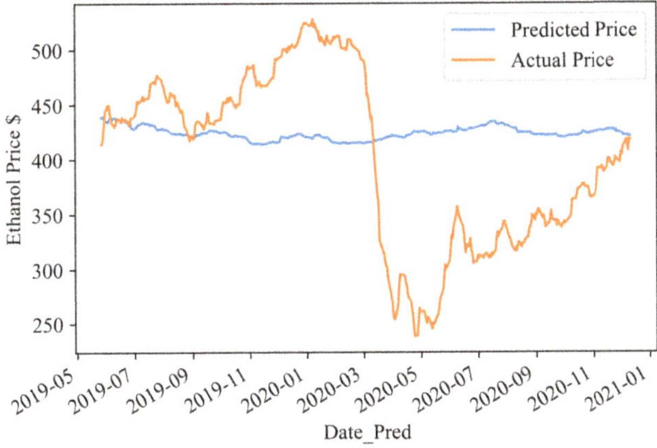

Figure 8. Predictions for 126 days ahead.

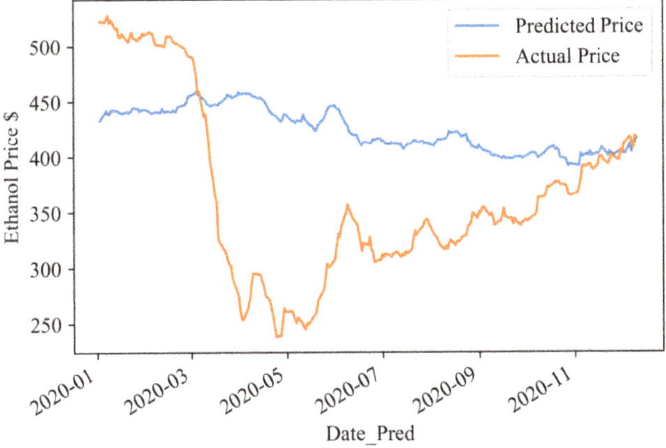

Figure 9. Predictions for 252 days ahead.

Furthermore, we can see that 63-days forecasts are relatively close to the actual price before the most intense period of the pandemic (more specifically between March and September of 2020) and, after that, recover similar performance. Thus, regarding regression outlines, there are evidences that incremental volatility could negatively interfere in the predictions. Besides, it is hard to conclude for the price directions, as we found accuracy higher than 80% for all the best predictors. The precision achieves values over 70%, except for uptrend in 252-days, in which the test set is heavily unbalanced (only 37 upwards against 200 downwards). Since the interval of time for predicting is relatively short (less than a year), these findings would be useful and sufficiently interesting for people who need or want to trade ethanol, whatever is the interest: hedge or speculation positions.

5. Conclusions

This paper presents a Brazilian spot ethanol price prediction model using artificial neural networks with the LSTM architecture and compared it to SVM and RF forecasts. The algorithms provide outlines for periods of 63, 126 and 252 business days. The results evaluated in this work show that it was possible to predict prices with a reasonable degree of accuracy in market directions for all horizons used.

Tests to verify overfitting were performed using learning curves, and the models converged in a satisfactory way, demonstrating a good fit of the neural network. Benchmark results show that LSTM produced the smallest regression errors (MAPE and RMSE). However, regarding the correctness of the direction in the predictions, other algorithms had better accuracy for specific horizons.

SVML proved to be the best algorithm for detecting trends achieving good results for all forecast windows used. Still, LSTM also managed to achieve satisfactory results for all forecasts, unlike RF and SVMR which had poor results for the 252-day horizon.

It was possible to observe in the LSTM outcomes an increase in the accuracy of the algorithms in longer forecast horizons, 72%, 74%, and 80% for 63, 126, and 252-day horizons, respectively. However, the mean absolute percentage error (MAPE) of the forecasts increased: 17.2%, 19.9% and 26.1% for 3, 6 and 12 months respectively. The same was found in the RMSE outputs. Furthermore, it is important to note that the COVID-19 pandemic caused an unexpected drop in prices, increasing model errors.

The high degree of correctness of models in the direction of prices can be useful in the development of new hedging strategies for market agents. In addition, it can help producers and cooperatives to protect their capital through planning that takes into account these forecasts.

This work contributes by demonstrating that LSTM networks are able to perform efficiently when predicting ethanol prices, a biofuel widely used in Brazil and worldwide, which has the capacity to replace fossil energy sources.

Nevertheless, this paper has limitations: (i) despite of the satisfactory results, we built models based on pure techniques, (ii) our research takes into account one single commodity with prices traded in one country but these data can be exclusively found in Brazil (Top 3 producer in the world) and there is no global market with standardized price for ethanol, (iii) potential effects (e.g., macroeconomic indicators) are not considered. However, the proposed (and best) model requires only historical values, and (iv) comparison with previous results was impracticable since data, performance measures and horizons did not exist in the literature, what shows the pioneering of this study.

For future work, we can add hybrid models that mix different network architectures and machine learning algorithms, such as Empirical Mode Decomposition. Thus, new features can be implemented, such as endogenous variables (technical indicators) and exogenous variables (exchange rate, inflation and prices of other commodities).

Author Contributions: Conceptualization, G.C.S., F.B. and A.C.P.V.; methodology, G.C.S. and F.B.; software, G.C.S. and M.F.S.; validation, F.B. and A.C.P.V.; formal analysis, A.C.P.V.; investigation, G.C.S.; resources, A.C.P.V.; data curation, M.F.S.; writing—original draft preparation, G.C.S.; writing—review and editing, F.B. and A.C.P.V.; visualization, G.C.S. and M.F.S.; supervision, A.C.P.V.; project administration, F.B.; funding acquisition, G.C.S. and F.B. All authors have read and agreed to the published version of the manuscript.

Funding: This research was funded by Sapiens Agro, National Council for Scientific and Technological Development (CNPq, grant number 438314/2018-2), Postgraduate Program in Electrical Engineering of the Elec. Eng. School (PPGEELT–FEELT–UFU) and Postgraduate and Research Board of the Federal University of Uberlândia (PROPP–UFU, process number 23117.079268/2021-03).

Institutional Review Board Statement: Not applicable.

Informed Consent Statement: Not applicable.

Data Availability Statement: Publicly available datasets were analyzed in this study. This data can be found here: https://www.cepea.esalq.usp.br/en/indicator/ethanol.aspx (accessed on 11 November 2021). The predictions generated by the LSTM models can be accessed in the following repository: https://github.com/gustavocavsantos/Ethanol-Price-Predictions (accessed on 11 November 2021).

Conflicts of Interest: The authors declare no conflict of interest.

References

1. Goldemberg, J. The ethanol program in Brazil. *Environ. Res. Lett.* **2006**, *1*, 014008. [CrossRef]
2. Silva, G.P.D.; Araújo, E.F.D.; Silva, D.O.; Guimarães, W.V. Ethanolic fermentation of sucrose, sugarcane juice and molasses by Escherichia coli strain KO11 and Klebsiella oxytoca strain P2. *Braz. J. Microbiol.* **2005**, *36*, 395–404. [CrossRef]
3. Lopes, M.L.; de Lima Paulillo, S.C.; Godoy, A.; Cherubin, R.A.; Lorenzi, M.S.; Giometti, F.H.C.; Bernardino, C.D.; de Amorim Neto, H.B.; de Amorim, H.V. Ethanol production in Brazil: A bridge between science and industry. *Braz. J. Microbiol.* **2016**, *47*, 64–76. [CrossRef] [PubMed]
4. Ministry of Agriculture, Fisheries and Supply—Ethanol Archives. Available online: https://www.gov.br/agricultura/pt-br/assuntos/sustentabilidade/agroenergia/arquivos-etanol-comercio-exterior-brasileiro/ (accessed on 11 November 2021).
5. Carpio, L.G.T. The effects of oil price volatility on ethanol, gasoline, and sugar price forecasts. *Energy* **2019**, *181*, 1012–1022. [CrossRef]
6. EIA—Today in Energy. Available online: https://www.eia.gov/todayinenergy/detail.php?id=47956 (accessed on 11 October 2021).
7. Hira, A.; de Oliveira, L.G. No substitute for oil? How Brazil developed its ethanol industry. *Energy Policy* **2009**, *37*, 2450–2456. [CrossRef]
8. David, S.A.; Inácio, C.; Tenreiro Machado, J.A. Quantifying the predictability and efficiency of the cointegrated ethanol and agricultural commodities price series. *Appl. Sci.* **2019**, *9*, 5303. [CrossRef]
9. David, S.; Quintino, D.; Inacio, C.; Machado, J. Fractional dynamic behavior in ethanol prices series. *J. Comput. Appl. Math.* **2018**, *339*, 85–93. [CrossRef]
10. Tapia Carpio, L.G.; Simone de Souza, F. Competition between second-generation ethanol and bioelectricity using the residual biomass of sugarcane: Effects of uncertainty on the production mix. *Molecules* **2019**, *24*, 369. [CrossRef] [PubMed]
11. Herrera, G.P.; Constantino, M.; Tabak, B.M.; Pistori, H.; Su, J.J.; Naranpanawa, A. Long-term forecast of energy commodities price using machine learning. *Energy* **2019**, *179*, 214–221. [CrossRef]
12. de Araujo, F.H.A.; Bejan, L.; Rosso, O.A.; Stosic, T. Permutation entropy and statistical complexity analysis of Brazilian agricultural commodities. *Entropy* **2019**, *21*, 1220. [CrossRef]
13. Barboza, F.; Kimura, H.; Altman, E. Machine learning models and bankruptcy prediction. *Expert Syst. Appl.* **2017**, *83*, 405–417. [CrossRef]
14. Alameer, Z.; Fathalla, A.; Li, K.; Ye, H.; Jianhua, Z. Multistep-ahead forecasting of coal prices using a hybrid deep learning model. *Resour. Policy* **2020**, *65*, 101588. [CrossRef]
15. Sun, W.; Zhang, J. Carbon Price Prediction Based on Ensemble Empirical Mode Decomposition and Extreme Learning Machine Optimized by Improved Bat Algorithm Considering Energy Price Factors. *Energies* **2020**, *13*, 3471. [CrossRef]
16. Bouri, E.; Dutta, A.; Saeed, T. Forecasting ethanol price volatility under structural breaks. *Biofuels Bioprod. Biorefining* **2021**, *15*, 250–256. [CrossRef]
17. Pokrivčák, J.; Rajčaniová, M. Crude oil price variability and its impact on ethanol prices. *Agric. Econ.* **2011**, *57*, 394–403. [CrossRef]
18. Bastianin, A.; Galeotti, M.; Manera, M. Ethanol and field crops: Is there a price connection? *Food Policy* **2016**, *63*, 53 – 61. [CrossRef]
19. Bildirici, M.; Guler Bayazit, N.; Ucan, Y. Analyzing Crude Oil Prices under the Impact of COVID-19 by Using LSTARGARCHLSTM. *Energies* **2020**, *13*, 2980. [CrossRef]

20. Ding, S.; Zhang, Y. Cross market predictions for commodity prices. *Econ. Model.* **2020**, *91*, 455–462. [CrossRef]
21. Kulkarni, S.; Haidar, I. Forecasting model for crude oil price using artificial neural networks and commodity futures prices. *arXiv* **2009**, arXiv:0906.4838.
22. Liu, Y.; Yang, C.; Huang, K.; Gui, W. Non-ferrous metals price forecasting based on variational mode decomposition and LSTM network. *Knowl.-Based Syst.* **2020**, *188*, 105006. [CrossRef]
23. Hu, Y.; Ni, J.; Wen, L. A hybrid deep learning approach by integrating LSTM-ANN networks with GARCH model for copper price volatility prediction. *Phys. A Stat. Mech. Its Appl.* **2020**, *557*, 124907. [CrossRef]
24. Zhou, B.; Zhao, S.; Chen, L.; Li, S.; Wu, Z.; Pan, G. Forecasting Price Trend of Bulk Commodities Leveraging Cross-Domain Open Data Fusion. *ACM Trans. Intell. Syst. Technol.* **2020**, *11*, 1–26. [CrossRef]
25. Ouyang, H.; Wei, X.; Wu, Q. Agricultural commodity futures prices prediction via long- and short-term time series network. *J. Appl. Econ.* **2019**, *22*, 468–483. [CrossRef]
26. CEPEA—Center for Advanced Studies on Applied Economics. Available online: https://www.cepea.esalq.usp.br/en/cepea-1.aspx (accessed on 13 January 2021).
27. Ariyawansha, T.; Abeyrathna, D.; Kulasekara, B.; Pottawela, D.; Kodithuwakku, D.; Ariyawansha, S.; Sewwandi, N.; Bandara, W.; Ahamed, T.; Noguchi, R. A novel approach to minimize energy requirements and maximize biomass utilization of the sugarcane harvesting system in Sri Lanka. *Energies* **2020**, *13*, 1497. [CrossRef]
28. Franken, J.R.; Parcell, J.L. Cash Ethanol Cross-Hedging Opportunities. *J. Agric. Appl. Econ.* **2003**, *35*, 510–516. [CrossRef]
29. Uhrig, R.E. Introduction to artificial neural networks. In Proceedings of IECON'95-21st Annual Conference on IEEE Industrial Electronics, Orlando, FL, USA, 6–10 November 1995; Volume 1, pp. 33–37.
30. Nakisa, B.; Rastgoo, M.N.; Rakotonirainy, A.; Maire, F.; Chandran, V. Long Short Term Memory Hyperparameter Optimization for a Neural Network Based Emotion Recognition Framework. *IEEE Access* **2018**, *6*, 49325–49338. [CrossRef]
31. Breuel, T.M. Benchmarking of LSTM Networks. *arXiv* **2015**, arXiv:1508.02774.
32. Hochreiter, S.; Schmidhuber, J. Long Short-Term Memory. *Neural Comput.* **1997**, *9*, 1735–1780. [CrossRef]
33. Gers, F.A.; Schmidhuber, J.; Cummins, F. Learning to Forget: Continual Prediction with LSTM. *Neural Comput.* **2000**, *12*, 2451–2471. [CrossRef]
34. Yu, Y.; Si, X.; Hu, C.; Zhang, J. A Review of Recurrent Neural Networks: LSTM Cells and Network Architectures. *Neural Comput.* **2019**, *31*, 1235–1270. [CrossRef]
35. Greff, K.; Srivastava, R.K.; Koutník, J.; Steunebrink, B.R.; Schmidhuber, J. LSTM: A Search Space Odyssey. *IEEE Trans. Neural Netw. Learn. Syst.* **2017**, *28*, 2222–2232. [CrossRef] [PubMed]
36. Graves, A.; Fernández, S.; Schmidhuber, J. Bidirectional LSTM Networks for Improved Phoneme Classification and Recognition. In *Artificial Neural Networks: Formal Models and Their Applications—ICANN 2005*; Duch, W., Kacprzyk, J., Oja, E., Zadrożny, S., Eds.; Springer: Berlin/Heidelberg, Germany, 2005; pp. 799–804.
37. Habibi, M.; Weber, L.; Neves, M.; Wiegandt, D.L.; Leser, U. Deep learning with word embeddings improves biomedical named entity recognition. *Bioinformatics* **2017**, *33*, i37–i48. [CrossRef] [PubMed]
38. Zhou, C.; Sun, C.; Liu, Z.; Lau, F. A C-LSTM neural network for text classification. *arXiv* **2015**, arXiv:1511.08630.
39. Zhou, P.; Qi, Z.; Zheng, S.; Xu, J.; Bao, H.; Xu, B. Text classification improved by integrating bidirectional LSTM with two-dimensional max pooling. *arXiv* **2016**, arXiv:1611.06639.
40. Liu, G.; Guo, J. Bidirectional LSTM with attention mechanism and convolutional layer for text classification. *Neurocomputing* **2019**, *337*, 325–338. [CrossRef]
41. Sachan, D.S.; Zaheer, M.; Salakhutdinov, R. Revisiting lstm networks for semi-supervised text classification via mixed objective function. In Proceedings of the AAAI Conference on Artificial Intelligence, Honolulu, HI, USA, 27 January–1 February 2019; Volume 33, pp. 6940–6948.
42. Bao, W.; Yue, J.; Rao, Y. A deep learning framework for financial time series using stacked autoencoders and long-short term memory. *PLoS ONE* **2017**, *12*, e0180944. [CrossRef]
43. Siami-Namini, S.; Tavakoli, N.; Namin, A.S. A comparison of ARIMA and LSTM in forecasting time series. In Proceedings of the 2018 17th IEEE International Conference on Machine Learning and Applications (ICMLA), Orlando, FL, USA, 17–20 December 2018; pp. 1394–1401.
44. Moghar, A.; Hamiche, M. Stock market prediction using LSTM recurrent neural network. *Procedia Comput. Sci.* **2020**, *170*, 1168–1173. [CrossRef]
45. Tong, G.; Yin, Z. Adaptive Trading System of Assets for International Cooperation in Agricultural Finance Based on Neural Network. *Comput. Econ.* **2021**, 1–20. [CrossRef]
46. Karim, F.; Majumdar, S.; Darabi, H.; Chen, S. LSTM Fully Convolutional Networks for Time Series Classification. *IEEE Access* **2018**, *6*, 1662–1669. [CrossRef]
47. Mahasseni, B.; Lam, M.; Todorovic, S. Unsupervised video summarization with adversarial lstm networks. In Proceedings of the IEEE conference on Computer Vision and Pattern Recognition, Honolulu, HI, USA, 21–26 July 2017; pp. 202–211.
48. Sønderby, S.K.; Sønderby, C.K.; Nielsen, H.; Winther, O. Convolutional LSTM networks for subcellular localization of proteins. In *International Conference on Algorithms for Computational Biology*; Springer: Cham, Switzerland, 2015; pp. 68–80.

49. Trinh, H.D.; Giupponi, L.; Dini, P. Mobile traffic prediction from raw data using LSTM networks. In Proceedings of the 2018 IEEE 29th Annual International Symposium on Personal, Indoor and Mobile Radio Communications (PIMRC), Bologna, Italy, 9–12 September 2018; pp. 1827–1832.

50. Ycart, A.; Benetos, E. A Study on LSTM Networks for Polyphonic Music Sequence Modelling. In Proceedings of the International Society for Music Information Retrieval Conference (ISMIR), Suzhou, China, 23–28 October 2017.

51. Srivastava, N.; Hinton, G.; Krizhevsky, A.; Sutskever, I.; Salakhutdinov, R. Dropout: A simple way to prevent neural networks from overfitting. *J. Mach. Learn. Res.* **2014**, *15*, 1929–1958.

52. Gers, F.A.; Schmidhuber, E. LSTM recurrent networks learn simple context-free and context-sensitive languages. *IEEE Trans. Neural Netw.* **2001**, *12*, 1333–1340. [CrossRef]

53. Herrera, G.P.; Constantino, M.; Tabak, B.M.; Pistori, H.; Su, J.J.; Naranpanawa, A. Data on forecasting energy prices using machine learning. *Data Brief* **2019**, *25*, 104122. [CrossRef] [PubMed]

54. Carrasco, M.; López, J.; Maldonado, S. Epsilon-nonparallel support vector regression. *Appl. Intell.* **2019**, *49*, 4223–4236. [CrossRef]

55. Wang, X.; Zhou, T.; Wang, X.; Fang, Y. Harshness-aware sentiment mining framework for product review. *Expert Syst. Appl.* **2022**, *187*, 115887. [CrossRef]

56. Zhang, P.; Yin, Z.Y.; Zheng, Y.; Gao, F.P. A LSTM surrogate modelling approach for caisson foundations. *Ocean Eng.* **2020**, *204*, 107263. [CrossRef]

Article

Energy Management Model for HVAC Control Supported by Reinforcement Learning

Pedro Macieira, Luis Gomes and Zita Vale *

GECAD—Research Group on Intelligent Engineering and Computing for Advanced Innovation and Development, Polytechnic of Porto (P.PORTO), P-4200-072 Porto, Portugal; picma@isep.ipp.pt (P.M.); lfg@isep.ipp.pt (L.G.)
* Correspondence: zav@isep.ipp.pt

Abstract: Heating, ventilating, and air conditioning (HVAC) units account for a significant consumption share in buildings, namely office buildings. Therefore, this paper addresses the possibility of having an intelligent and more cost-effective solution for the management of HVAC units in office buildings. The method applied in this paper divides the addressed problem into three steps: (i) the continuous acquisition of data provided by an open-source building energy management systems, (ii) the proposed learning and predictive model able to predict if users will be working in a given location, and (iii) the proposed decision model to manage the HVAC units according to the prediction of users, current environmental context, and current energy prices. The results show that the proposed predictive model was able to achieve a 93.8% accuracy and that the proposed decision tree enabled the maintenance of users' comfort. The results demonstrate that the proposed solution is able to run in real-time in a real office building, making it a possible solution for smart buildings.

Keywords: building energy management systems; HVAC control; Internet of Things; occupancy prediction; reinforcement learning

Citation: Macieira, P.; Gomes, L.; Vale, Z. Energy Management Model for HVAC Control Supported by Reinforcement Learning. *Energies* **2021**, *14*, 8210. https://doi.org/10.3390/en14248210

Academic Editors: Ana-Belén Gil-González and Fabrizio Ascione

Received: 26 October 2021
Accepted: 3 December 2021
Published: 7 December 2021

Publisher's Note: MDPI stays neutral with regard to jurisdictional claims in published maps and institutional affiliations.

1. Introduction

Smart grids enable the active participation of end-users and allow them to actively manage their demand using demand-side management [1,2]. By managing home available energy resources and loads, the end-users are able to reduce energy costs [3], maximize the use of renewable generation [4], participate in the smart grid [5], and transact energy with other end-users [6].

The ability to manage their own resources enables end-users to take control over their facilities and promote the dissemination of smart buildings [7]. Actions for demand-side management and for the active participation of end-users can also be achieved using internet of things (IoT) devices [8]. The possibilities of energy management are vast and they represent a good opportunity for end-users to enable the reduction of energy costs and optimization of energy usage [9].

One of the biggest contributions to the energy consumption in buildings comes from heating ventilating air conditioning (HVAC) systems [10]. Therefore, the control of HVAC systems is frequently found in the literature as a way to reduce energy costs and greenhouse gas emissions [11,12]. The control of HVAC can also be used for the active participation of end-users in smart grids [13]. Because of their characteristic preconditioning time, i.e., the preheating or cooling time of HVAC, the existing models for HVAC control usually results in planned action ahead in time, demanding the use of prediction models to forecast contexts. The use of prediction models, in smart grids, is common, in particular, to forecast the consumption of end-user and individual loads, the generation of renewable resources, and the flexibility of end-users [14].

The optimization of HVAC systems, as any other electrical load placed in the end-user facility, needs to consider the users' preferences and needs [15]. It is then necessary to balance the users' comfort and the energy cost reduction. This is where prioritization techniques come in, allowing energy consumption to be reduced and avoiding unnecessary electricity expenses [16].

This paper proposes a novel model for HVAC control using a prediction reinforcement learning algorithm and a decision tree for the consideration of building context, energy costs, and preconditioning time. The prediction model is used to predict occupancy in the building ahead in time. The occupancy prediction is then used in a decision tree to control of the HVAC units, considering the energy prices and the current temperature of the building. The proposed solution can intelligently control the HVAC units and promote the reduction of energy costs.

The proposed solution, comprised of the prediction model and the decision tree, was deployed in a real office building using a building energy management system (BEMS) that was implemented based on the open-source platform Home Assistant (https://www. home-assistant.io/, accessed on 6 December 2021). This BEMS enabled the integration of multiple IoT devices and the implementation of the proposed models in a real building, allowing a continuous operation. The proposed solution addresses some of the limitations that were identified in the literature review, such as continuous learning, the consideration of several contexts in the building, and the users' comfort.

This paper is structured as follows. After this first introductory section, related works regarding the management of HVAC systems, namely occupancy-based solutions, are presented in Section 2. In Section 3, several solutions to develop a BEMS are presented and the proposed BEMS, based on Home Assistant, is described. Section 4 presents the proposed methodology including the prediction model and the decision tree. The results of the case study, using a real building, are described in Section 5. Section 6 presents the discussion of results, while the main conclusions are presented in Section 6.

2. Related Works

The occupancy of the building and/or of different building zones can be used by energy management systems to promote the contextual optimization of resources [15]. The information regarding occupancy allows systems to better understand the building context and provide better resource optimization [17]. The exact occupation of the building and the location of people is hard to achieve, but estimates can be obtained using several techniques, using equipment already installed in the building or that requires new equipment to be installed [18]. In [19], a machine learning classification model is proposed for week-ahead occupancy prediction using real-time smart meter data. Applied to HVAC systems, [20] proposed a control model based on students' location inside auditoriums. This model uses Wi-Fi data to determine the number of connected devices and enables a noninvasive approach that does not require the installation of new equipment. In [21], a model for HVAC control inside buildings is proposed using a rule-based approach supported by a deep learning algorithm that predicts the preconditioning time. These solutions address the prediction and identification of occupancy in buildings. However, they lack the ability to consider multiple contexts. The contextual reinforcement learning predictive model proposed in this paper addresses the dynamics of the building usage considering several contexts.

A reinforcement learning model for occupant behavior is proposed in [22] using a Q-learning model to allow automated control of the building's thermostat. This work compared the prediction of the reinforcement learning model with an artificial neural network (ANN) prediction model. The results showed that the ANN had better results. In [23] the use of reinforcement learning models to directly control the HVAC units demonstrated an increase in energy costs, even when dealing with on/off control. However, the best results were achieved when the model was used to control five flow levels.

Other contextual data, beyond occupancy, can be used to control HVAC systems [24]. A model for HVAC control considering not only the building occupancy but also the characterization of the building, namely the mean radiant temperature of building rooms is proposed in [25]. In [26], it is proposed the use of IoT devices to obtain contextual information regarding outdoor temperature to improve the efficiency of HVAC control and achieve lower payback times, when compared to occupancy-based models. However, these solutions lack the ability to learn from historical data, and more importantly, the ability to be able to continuously learn. The proposed solution addresses this limitation by conceiving a reinforcement learning model.

The control and management of HVAC systems are usually used in the energy domain as a way to reduce energy costs [27], due to the high energy consumption of HVAC equipment. A day-ahead HVAC control model for industrial buildings, to decrease energy costs considering on-peak and off-peak tariffs and weather conditions, is proposed in [28]. A multiperiod optimization model for supervisory control and data acquisition (SCADA) systems is proposed in [29] to minimize energy costs while complying with users constraints. In [30], HVAC systems are controlled using market-based transactive controls in commercial buildings to promote grid balance and stability. Although energy costs are an important aspect of the building, the user comfort needs to be considered, as well as the building's context. The proposed solution addresses the users' comfort considering ahead HVAC control signals to prepare the building for the users' arrival while trying to minimize energy costs. The context of the building is considered in the proposed solution, for instance, the HVAC units' control considers the status of the windows.

The control of HVAC units can also be done to improve air quality. In [31], a computer vision-based system is used to get the occupancy of a room and a neural network is used to classify the users' activities. This solution enables the prediction of CO_2 levels and controls the HVAC unit to prevent potentially dangerous situations. In [32], a solution for commercial buildings is proposed where a scalable model, based on multiagent deep reinforcement learning, is used to minimize energy costs while considering user comfort and air quality.

More complex solutions can also be implemented to enable the combined management of multi-HVAC systems. In [33], a data-driven mathematical model using contextual sensor data is used to optimize multi-HVAC systems that serve the same space. In [34], several optimization models are compared for multi-HVAC systems management considering the outdoor temperature, building ambient temperature, and occupancy. The compared models were a multiobjective genetic algorithm, two nondominated sorting genetic algorithms, an optimized multiobjective particle swarm optimization, a speed-constrained multi-objective particle swarm optimization, and a random search. The authors were not able to identify a unique best solution but presented the benefits of each model.

3. Building Energy Management System Based on Home Assistant

A BEMS can be installed in a variety of ways, normally this is done using dedicated, commercial platforms such as DEXMA Energy Intelligence (https://www.dexma.com/, accessed on 6 December 2021) or Entronix (https://entronix.io/, accessed on 6 December 2021). However, for the purposes of this paper, a home automation system was used, enabling the combination of Internet of Things (IoT) devices with the ability to gather data and affect the environment around them with rules created by the user to manage the smart home.

There are a variety of home automation system software solutions, both open and closed source. The biggest open-source solutions available are:

- openHAB (https://www.openhab.org/, accessed on 6 December 2021), created by Kai Kreuzer, which is built using Java and was initially released in 2010;
- Home Assistant, created by Paulus Schoutsen, based on Python and configurable with the use of YAML files, which was first released in 2013;
- Domoticz (https://www.domoticz.com/, accessed on 6 December 2021), created by Gizmocuz, is written in C++ and configurable with Blockly or Lua, and with the first release date at the end of 2012.

There are also a wide variety of commercial closed source solutions, with the biggest ones being owned by large players:

- Google Assistant (https://assistant.google.com/, accessed on 6 December 2021), Google's solution for home assistant software, allows the use of Google Assistant software together with Google Home devices to control a house with voice commands and using a graphical user interface (GUI). It also enables the customization of automation rules;
- Alexa (https://www.amazon.com/b?ie=UTF8&node=21576558011, accessed on 6 December 2021), Amazon's solution for home automation, works similarly to Google Assistant, by using an Alexa device to allow the user to control their smart home with voice commands and with GUI;
- HomeKit (https://www.apple.com/ios/home/, accessed on 6 December 2021), developed by Apple, allows users to integrate their houses with the Apple Ecosystem, allowing the use of voice commands through Siri but also providing a GUI for more powerful smart home management.

For the purposes of this paper, Home Assistant was chosen due to its many advantages such as being open-source, easily customizable with simple YAML files, and the powerful but still simple set-up GUI.

First, some IoT devices were installed and configured in a docker containerized Home Assistant installation, with their uses described in Table 1. The IoT devices integrated can be seen as actuators and sensors, depending on the type of actions they allow. Additionally to market-available IoT devices, the Home Assistant configuration also integrates a SCADA system, developed by the researchers of the authors' research center, that was already present in the building. This SCADA system is represented in Table 1 as GECAD API. This API enables the reading of other IoT devices that are already integrated into the SCADA system by using REST-based HTTP requests. The last device identified in Table 1 is a simulated sensor that was integrated into Home Assistant as a file sensor, enabling the reading of a file and the publishing of that information. The file sensor is a dataset of real electricity prices that were obtained from MIBEL, the Iberian electricity wholesale market.

Table 1. Devices configured for the Home Assistant installation.

Device Name	Monitoring Data	Actions
RM Pro +	Any IR or RF signal	Emission of IR or RF signal
Sonoff Pow R2	Current (A), power (W), voltage (V), load status (on/off)	Turn on/off
D-Link DSP-W115	Load status (on/off)	Turn on/off
D-Link DSP-W215	Power (W), total power consumption (kWh), temperature (°C)	Turn on/off
Sonoff RF Bridge	N/A	N/A
Sonoff Sensor DW1	Door/window open	triggered/A
GECAD API (REST Sensors)	Exterior temperature (°C) Interior temperature of individual rooms (°C) Light switch states (%)	N/A
File Sensor (JSON Source)	Electricity price on a specific month, time, day, and hour (EUR/kWh)	N/A

To manage the energy of a building in an intelligent way using Home Assistant, complex logic needs to be integrated. However, Home Assistant has some constraints, namely the fact that it is configured by YAML files, which can turn simple scripts into a large amount of complex virtual sensors that are tightly interconnected and have little use outside of single automation. This difficult the setting-up of the installation and makes it very hard to share the logic with other Home Assistant installations.

To solve this issue, Pyscript (https://github.com/custom-components/pyscript, accessed on 6 December 2021), an add-on to Home Assistant, has been used for the work presented in this paper. This add-on can import standard Python libraries and interact with Home Assistant triggers and events when IoT devices or Home Assistant itself create them. In this way, the user can create applications that can be configured with YAML when there is a need to replicate them multiple times, and, most importantly, it allows the use of Python to write the scripts that can interact in real-time with Home Assistant. This enables the Home Assistant to be integrated with all kinds of external services that can manage its resources.

Home Assistant was configured with ten views, which resulted in tabs inside the Home Assistant dashboard, one for each building zone, and a general view with data that is not specific to any zone. The building is divided into nine zones, where each has from two to three rooms that can be offices, meetings rooms, server rooms, or laboratories. Figure 1 presents the aspect of the interface for zone 1 of the considered building.

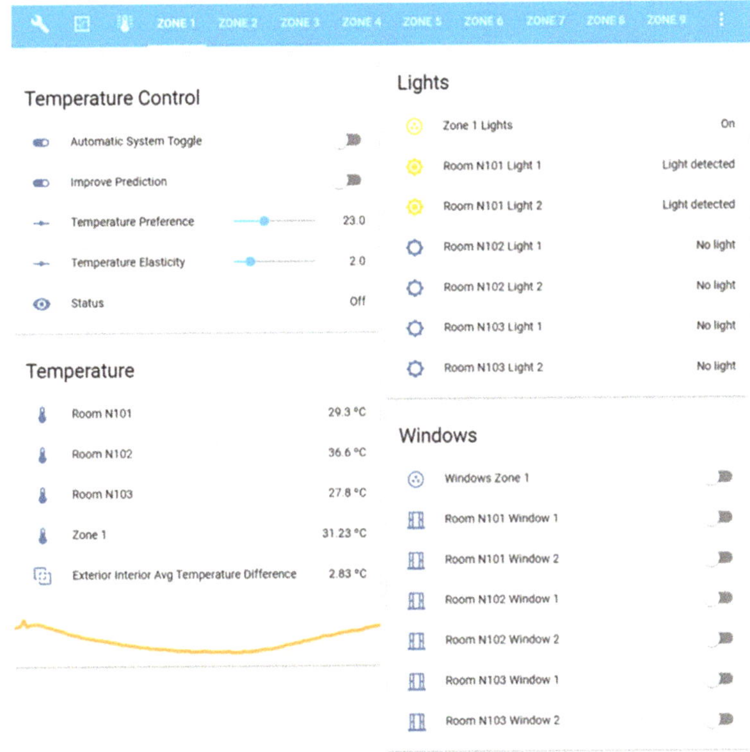

Figure 1. The Home Assistant interface for Zone 1.

The view for each zone contains multiple panels for IoT device monitoring and control. It has a 'temperature control' panel where the users can activate and control the proposed HVAC control model, allowing:

- The activation of the proposed HVAC control model;
- The activation of continuous learning, to enable the reinforcement learning prediction model to learn;
- The setting of user temperature preferences;
- The setting of user temperature elasticity;
- The current status of the system (e.g., "On, heating up" or "Off" or "Off, the window is open").

The general view of the Home Assistant interface can be seen in Figure 2 where electricity market-related data are presented. In the general view, it is also possible to control the maximum price for the system to run autonomously, to save money, and avoid turning on the energy-expensive HVAC systems when energy costs are too high.

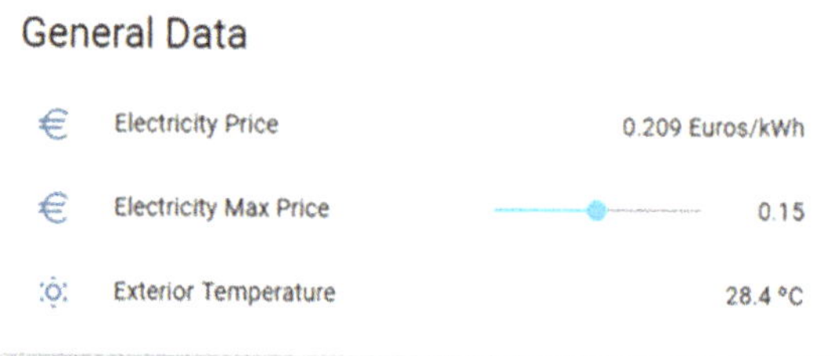

Figure 2. The Home Assistant interface for general data.

4. Method and Proposed Solution

Currently, most HVAC systems are manually controlled by users who are on-site and find themselves at an uncomfortable temperature level. The user must manually configure the HVAC system and wait for it to effectively adjust the temperature up or down to a more comfortable level. Alternatively, there is a central system that can be configured with sensors to detect temperature changes and automatically control the HVAC system. However, central systems also raise some issues, such as the waiting time until the HVAC system successfully regulates itself back to a comfortable temperature. An even bigger issue is the fact that it may be turning the HVAC system on when there are no users in the room/building, thus resulting in the waste of energy.

The proposed solution considers three components that will enable the individual intelligent management of each HVAC unit present in the building. The first component is the deployment of the open-source BEMS that is presented in Section 3. The BEMS enables the continuous monitoring and control of the building's resources and the storing of the data. The historical data of the BEMS is fed to a predictive model supported by machine learning. The predictive model is proposed to enable the prediction of office space usage. The last component of the proposed solution is the proposed decision tree that will act on the HVAC units considering the prediction, current environmental context, the limits and preferences of the users, and the current energy prices.

Figure 3 shows the complete flowchart of the proposed solution. The flowchart represents three threads that are executed. The first one is executed in every instant to allow Home Assistant to have continuous monitoring of the building. The second one is used to improve the reinforcement learning algorithm used as a predictive model, and it is executed once every period. This flow monitors the room's occupancy, stores the data, and recalculates the rewards. In the last flow, the control over the HVAC units is performed. The last thread is executed once every period and is able to collect the energy price, predict the room's occupancy, apply the decision tree, and control the HVAC units accordingly.

This paper proposes the use of a contextual reinforcement learning algorithm to predict the occupancy of the rooms of a building. The use of reinforcement learning enables the proposed model to have a continuous learning process, suitable to adjust the prediction to new routines or the addition of team members working in the same office space. Moreover, the proposed model is a contextual reinforcement learning algorithm that is able to have predictions considering the context of the office.

The reinforcement learning algorithm's configuration and tuning are described and a test between linear and neural models is done to obtain the best learning accuracy. The paper also proposes a novel decision tree to enable the ahead control of HVAC units taking into consideration energy prices and user preferences, namely the desired temperature. This proposed methodology was integrated into the proposed building energy management system, presented in Section 3.

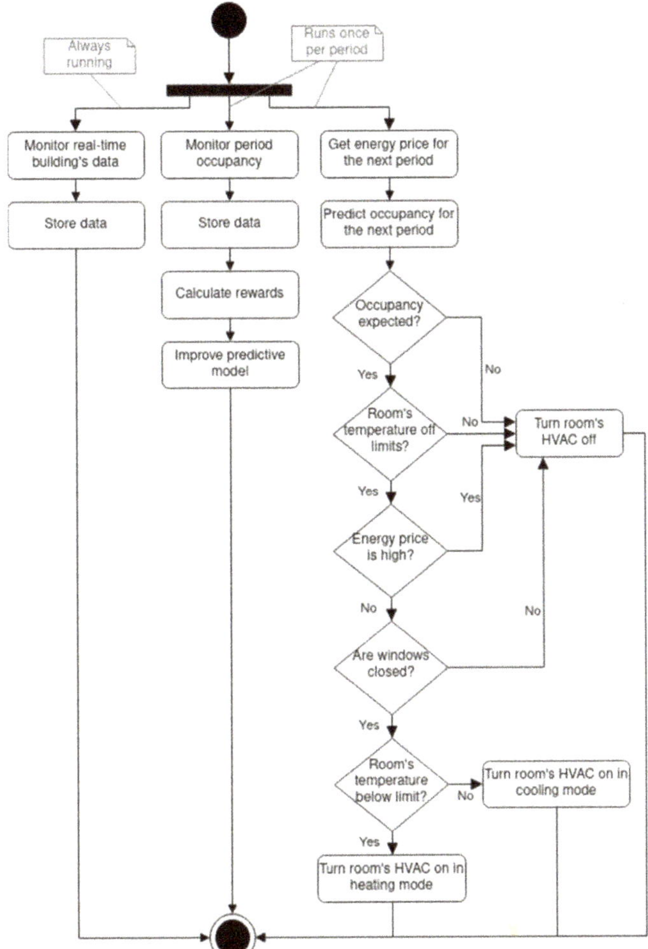

Figure 3. Solution flowchart.

4.1. Reinforcement Learning Model for Occupancy Forecast

The proposed HVAC control model is based on the reinforcement learning algorithm published in [35], enabling the use of contextual reinforcement learning, used in this paper to predict the building's occupancy. In this paper, the algorithm was configured and trained to enable queries, using a specific date and time, to return an occupancy prediction as a response. The combination of this occupancy model with the previously described home automation system will allow for the setup of an intelligent HVAC system management system capable of correctly setting the temperature while still avoiding the situation of energy waste.

The biggest advantage of this model is that when the model wrongfully predicts the occupancy of the building, it will also be capable of learning from its mistakes and adapting itself over time, eventually figuring out the new pattern of occupation. This enables the system to self-adjust itself to new realities of the building, such as the changing of usage patterns created by new collaborators.

The model is executed every 30 min, every day of the week in a continuous fashion. Every time it is queried it will attempt to predict if a group of rooms, henceforth called a zone, will be occupied or not on the next 30 min period, thus giving ample time for the HVAC to adjust the room temperature, if needed.

4.2. Hyperparameter Tuning

After the initial training of the model, it was further refined with a process of hyperparameter tuning. In this way, the model can achieve the highest possible accuracy score and the correct amount of learning/exploring so it can keep high accuracy scores in the future. For this, a specific set of parameters were tuned:

- The number of hidden layers of the neural model;
- The size of each hidden layer;
- The learning rate;
- The decay of the learning rate;
- A 0 and β 0 which are used to calculate the inverse γ and Gaussian inference used by the model to perform exploration;
- The λ prior.

4.3. Decision Tree

After the prediction of occupancy, a decision tree is used to identify the need for HVAC control. The decision tree is shown in Figure 3, being executed after the predictive model. There are three possible actions resulting from the decision tree: turn off the HVAC unit of the room, turn it on in cooling mode, and turn it on in heating mode.

The decision tree is used to control the HVAC system considering several contextual variables. The decision tree is applied 15 min ahead and during the targeted period, enabling a continuous monitoring and feedback loop between the context and the HVAC control. The decision tree was constructed to consider more parameters than just whether the zone will be occupied or not, taking into consideration:

- Current room temperature to check if there is even a need for the HVAC to be turned on in the first place;
- If there are open windows which would represent a significant waste of energy due to the outside–inside temperature differential;
- If the current electricity price is too high;
- Whether it will turn the heating on or the cooling on.

The variables used in the decision tree are a combination of sensor data and user specifications. The users of each office space are responsible to agree, converge, and define the temperature comfortable limits. The building's owner needs also to define the energy price threshold used for the decision tree.

5. Case Study

The proposed solution was deployed in a multiple office building with IoT sensors that were configured with the used home automation software (i.e., Home Assistant). The building is organized into nine zones. The proposed HVAC model was tested on Zone 1, which contains three office rooms named N101, N102, and N103. An overview of the floorplan of the building and an aerial photo (Figure 4) demonstrates how the building is laid out.

5.1. Training Dataset

The dataset used to train the learning model consists of one-year data collected in 2019. The decision to use the year 2019 data is due to the SARS-CoV-2 restrictions that have been in use since March 2020. In this way, during 2020 the building had usage patterns radically different from the ones occurring during a normal year. The structure of the dataset can be seen in Table 2.

Figure 4. Aerial view of the building and its floorplan.

Table 2. Breakdown of the preprocessed dataset.

Column	Description
Weekday	The day of the week as an integer (i.e., from 0 to 6)
Time	The time that marks the start of the interval in seconds counting from midnight (i.e., from 0 to 86,400)
Occupied	A Boolean representing whether the zone was occupied during the interval or not (i.e., true or false)

The dataset consists of 17,520 entries, each entry corresponding to a single 30-min interval of the year 2019. The occupation value was not directly provided by any IoT integrated into the proposed solution but has been inferred from the ceiling lamp consumption. If during the 30-min interval the power draw of any lamp in the zone was higher than 0 watts, the zone was assumed to be occupied, which is a sensible approach in face of the building usage patterns.

The initial dataset was then split into two. A subdataset of 16,176 entries (92.32%) was used for the training of the reinforcement learning model. The remaining 1344 entries (7.68%), representing 4 weeks, were used for evaluation. The split of the dataset was done by weekly periods, meaning that the evaluation subdataset has four weeks of random months, but where each counts with seven sequential days.

As the building is not usually open outside business hours, the majority of the entries have the building marked as unoccupied, meaning that the used dataset was unbalanced. A density graph of the occupation of the building elaborated from this dataset can be viewed in Figure 5.

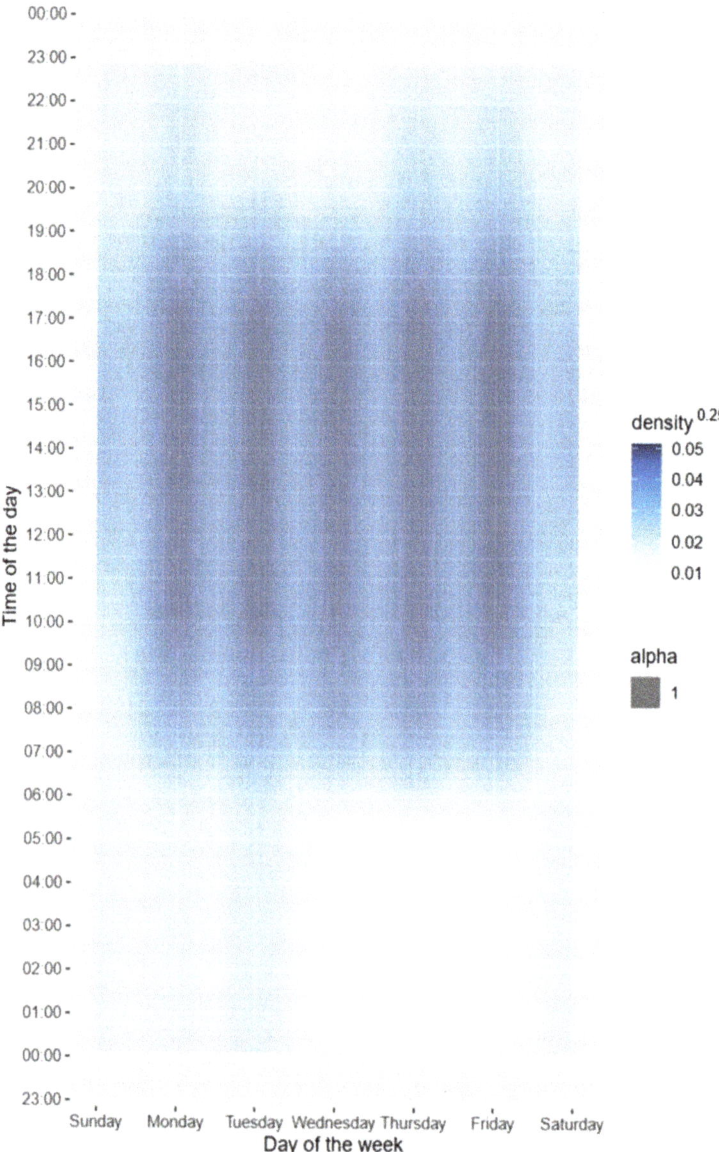

Figure 5. Occupancy graph of the building during 2019.

5.2. Forecast Errors Evaluation

Every time the model is executed, it takes in two input parameters, the current day of the week represented as an integer, ranging from 0 representing Monday to 6 representing Saturday, and the number of seconds that have passed since midnight of the current day up until the desired 30-min prediction interval.

Two different models were evaluated based on [36]: a linear model that simply attempted to regress the inputs and predict an output, and a more complex neural linear model that was able to learn a representation of the inputs and make a prediction based on that representation.

The training dataset was used on both reinforcement learning models to compare and evaluate their results. This evaluation step allowed the identification of the model that best performs under our case study.

The linear model was faster to train but it was limited by its ability to represent the problem and accurately predict whether the zone would be occupied or not, mostly predicting that it would be empty. Because an unbalanced dataset was used, the linear model managed to achieve an accuracy of around 68% as can be seen in its confusion matrix in Table 3. The neural linear model, using its default configuration of hyperparameters, was able to learn from the training dataset and achieve an accuracy of 92% (Table 4). From this step forward, the linear model was discarded due to its low accuracy score and the focus was set on improving the result of the neural linear model.

Table 3. Confusion matrix for the linear prediction model.

True Label	Predicted Label	
	Empty Zone	**Occupied Zone**
Empty zone	855	64
Occupied zone	373	80
Accuracy	68.15%	

Table 4. Confusion matrix for the neural linear prediction model.

True Label	Predicted Label	
	Empty Zone	**Occupied Zone**
Empty zone	873	46
Occupied zone	60	393
Accuracy	92.27%	

5.3. Hyperparameter Tuning Results

To tune the hyperparameters, Python's Optuna library was used due to its flexibility and learning curve, along with the ability to provide useful data and information at the end of a hyperparameter tuning session.

First, a selection of hyperparameters to tune was made, and then a range of appropriate values was chosen for each of the hyperparameters to be tuned. Then, using a loop, a random combination of hyperparameters was chosen and the model was trained and evaluated. To avoid wasting time in nonviable trials, the algorithm continuously checks the current accuracy and if it were lower than the average accuracy of the previous trials the entire trial would be pruned and the next trial starts.

In this case study, a hyperparameter tuning session with 500 trials was executed. From the obtained results, more sensible ranges were defined to achieve more accurate models. After another 400 trials, the most accurate model to come out of the hyperparameter tuning sessions was able to achieve an accuracy of 93.8%.

The configuration with the best result uses a network with three hidden layers with 26 nodes in the first and in the second layers, and with 12 nodes for the last layer. The chosen network had a learning rate of 0.03473 and a lambda of 0.20787.

5.4. HVAC Control Test

The control of HVAC units was done using Broadlink RM pro devices that were integrated into the Home Assistant solution. The Broadlink RM pro devices were located in each room of the building in Figure 4. The control of HVAC units was done, by the proposed solution, considering the decision tree shown in Figure 3. The decision tree was continuously executed in the proposed BEMS. However, the prediction value was only updated every 15 min and it considered the predicted presence of users for the next 15-min period. Therefore, the control was done ahead of time, i.e., to prepare the room for the next period by considering the time that HVAC units take to reach the desired temperature.

For this case study, the business day of Friday, 27 August 2021, was considered. Figure 6 shows the results for 24 h. The orange area represents the hours where users are predicted to be inside the office N101, part of Zone 1. The case study data considered two temperature limits of 25 and 23 °C, with a range of +3 °C, meaning that at the beginning of the day the temperature limit was from 25 °C to 28 °C, and it was changed to a limit between 23 °C and 26 °C. The energy threshold was set to 0.24 EUR/kWh.

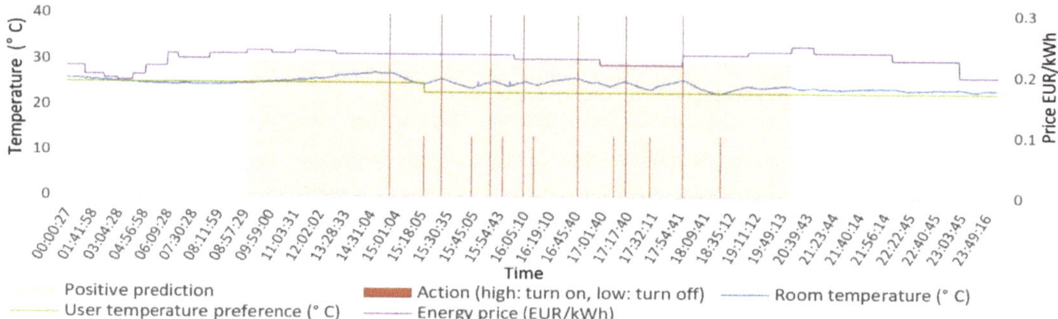

Figure 6. HVAC control signals for N101 during 27 August 2021.

6. Discussion

The results shown in Figure 6 indicate that the HVAC unit of room N101 was turned on seven times and turned off seven times over the 24 h. As can be seen, these actions happen during the hours where the predictive model forecasts the presence of users inside the office space of room N101, i.e., from 9:00 a.m. to 8:00 p.m. The turning on of the HVAC units also matched the increase of temperature above the upper user limit, while the turning off matched the below limit.

At 3:18 p.m., the user updated the temperature limit to a range between 23 °C and 26 °C, changing the system performance. This action tested the system's ability to adjust to the users' needs. However, as future improvement, the authors suggest the use of building thermal modeling to increase the intelligence of the management model.

The windows of room N101 were open between 4:07 p.m. and 4:45 p.m., leading to an early turn-off of the HVAC unit at 4:08 p.m. when the temperature was 24.93 °C (i.e., 1.93 °C above the lower limit of 23 °C). When the windows were closed, the HVAC almost immediately started once again to decrease the room's temperature.

The system demonstrated its ability to continuously manage the HVAC unit considering the current context of the room while also considering the predictive data of the proposed model. The control of the temperature inside the room was not only made according to the current context. If the predictive model did not forecast the presence of users during the next period, no control would be made in the HVAC units. All control signals, to turn the unit on, were made because the predictive model forecasted the presence of users during the next period.

The BEMS, based on the open-source solution of Home Assistant, was able to provide real-time monitoring and control while providing historical data to train the reinforcement learning algorithm. The proposed predictive model and the decision tree were deployed in BEMS using Python language and were able to cooperate with the BEMS to provide a continuous intelligent operation without the need for manual actions.

7. Conclusions

This paper proposes a novel model for the control of HVAC units based on a prediction model and a decision tree. The control of the units is performed ahead of time to minimize the units' waiting times (i.e., the time they take to reach the desired temperature). The prediction model uses a reinforcement algorithm that can provide continuous learning according to several contexts. The high accuracy of the model, 93.8%, demonstrates the ability of such algorithms to be used in smart buildings to efficiently manage energy loads and resources across different contexts.

Using the prediction result, a decision tree is proposed to consider the current context and the predicted context, inside a building's zone, to control the HVAC units. The proposed model was tested and evaluated using real data from one year in an office building. The validation of the proposed solution for 24 h is presented to demonstrate the use of the proposed solution as a whole.

Moreover, the work described in the paper demonstrated the ability to have open-source-based automation solutions to deploy complex models of artificial intelligence and energy management to operate in real-time and also ahead of time. This combination of technologies provides a complete tool to test and validate energy management models in real buildings.

Author Contributions: Conceptualization, P.M., L.G. and Z.V.; methodology, P.M., L.G. and Z.V.; software, P.M.; validation, P.M. and L.G.; formal analysis, P.M. and L.G.; investigation, P.M.; resources, Z.V.; data curation, P.M.; writing—original draft preparation, P.M. and L.G.; writing—review and editing, L.G. and Z.V.; visualization, P.M. and L.G.; supervision, L.G. and Z.V.; project administration, Z.V.; funding acquisition, Z.V. All authors have read and agreed to the published version of the manuscript.

Funding: The present work has received funding from European Regional Development Fund through COMPETE 2020—Operational Programme for Competitiveness and Internationalisation through the P2020 Project TIoCPS (ANI | P2020 POCI-01-0247-FEDER-046182), and has been developed under the EUREKA—ITEA3 Project TIoCPS (ITEA-18008).

Institutional Review Board Statement: Not Applicable.

Informed Consent Statement: Not Applicable.

Data Availability Statement: Not Applicable.

Acknowledgments: The authors acknowledge the work facilities and equipment provided by GECAD research center (UIDB/00760/2020, UIDP/00760/2020) to the project team.

Conflicts of Interest: The authors declare no conflict of interest.

References

1. Lezama, F.; Soares, J.; Canizes, B.; Vale, Z. Flexibility management model of home appliances to support DSO requests in smart grids. *Sustain. Cities Soc.* **2020**, *55*, 102048. [CrossRef]
2. Gazafroudi, A.S.; Soares, J.; Ghazvini, M.A.F.; Pinto, T.; Vale, Z.; Corchado, J.M. Stochastic interval-based optimal offering model for residential energy management systems by household owners. *Int. J. Electr. Power Energy Syst.* **2019**, *105*, 201–219. [CrossRef]
3. Yu, L.; Xie, W.; Xie, D.; Zou, Y.; Zhang, D.; Sun, Z.; Zhang, L.; Zhang, Y.; Jiang, T. Deep Reinforcement Learning for Smart Home Energy Management. *IEEE Internet Things J.* **2020**, *7*, 2751–2762. [CrossRef]
4. Liere-Netheler, I.; Schuldt, F.; von Maydell, K.; Agert, C. Simulation of Incidental Distributed Generation Curtailment to Maximize the Integration of Renewable Energy Generation in Power Systems. *Energies* **2020**, *13*, 4173. [CrossRef]
5. Pinto, T.; Morais, H.; Sousa, T.M.; Sousa, T.; Vale, Z.; Praça, I.; Faia, R.; Pires, E.J.S. Adaptive Portfolio Optimization for Multiple Electricity Markets Participation. *IEEE Trans. Neural Netw. Learn. Syst.* **2016**, *27*, 1720–1733. [CrossRef]
6. Gomes, L.; Vale, Z.A.; Corchado, J.M. Multi-Agent Microgrid Management System for Single-Board Computers: A Case Study on Peer-to-Peer Energy Trading. *IEEE Access* **2020**, *8*, 64169–64183. [CrossRef]
7. Yu, L.; Qin, S.; Zhang, M.; Shen, C.; Jiang, T.; Guan, X. Deep Reinforcement Learning for Smart Building Energy Management: A Survey. *arXiv* **2020**, arXiv:2008.05074.
8. Gomes, L.; Ramos, C.; Jozi, A.; Serra, B.; Paiva, L.; Vale, Z. IoH: A Platform for the Intelligence of Home with a Context Awareness and Ambient Intelligence Approach. *Futur. Internet* **2019**, *11*, 58. [CrossRef]

9. Djenouri, D.; Laidi, R.; Djenouri, Y.; Balasingham, I. Machine learning for smart building applications: Review and taxonomy. *ACM Comput. Surv.* **2019**, *52*, 1–36. [CrossRef]
10. Pérez-Lombard, L.; Ortiz, J.; Pout, C. A review on buildings energy consumption information. *Energy Build.* **2008**, *40*, 394–398. [CrossRef]
11. Esrafilian-Najafabadi, M.; Haghighat, F. Occupancy-based HVAC control systems in buildings: A state-of-the-art review. *Build. Environ.* **2021**, *197*, 107810. [CrossRef]
12. Gholamzadehmir, M.; Del Pero, C.; Buffa, S.; Fedrizzi, R.; Aste, N. Adaptive-predictive control strategy for HVAC systems in smart buildings—A review. *Sustain. Cities Soc.* **2020**, *63*, 102480. [CrossRef]
13. Adhikari, R.; Pipattanasomporn, M.; Rahman, S. An algorithm for optimal management of aggregated HVAC power demand using smart thermostats. *Appl. Energy* **2018**, *217*, 166–177. [CrossRef]
14. Khalid, R.; Javaid, N. A survey on hyperparameters optimization algorithms of forecasting models in smart grid. *Sustain. Cities Soc.* **2020**, *61*, 102275. [CrossRef]
15. Jung, W.; Jazizadeh, F. Human-in-the-loop HVAC operations: A quantitative review on occupancy, comfort, and energy-efficiency dimensions. *Appl. Energy* **2019**, *239*, 1471–1508. [CrossRef]
16. Gomes, L.; Spínola, J.; Vale, Z.; Corchado, J.M. Agent-based architecture for demand side management using real-time resources' priorities and a deterministic optimization algorithm. *J. Clean. Prod.* **2019**, *241*, 118154. [CrossRef]
17. Petrosanu, D.M.; Carutasu, G.; Carutasu, N.L.; Pîrjan, A. A review of the recent developments in integrating machine learning models with sensor devices in the smart buildings sector with a view to attaining enhanced sensing, energy efficiency, and optimal building management. *Energies* **2019**, *12*, 4745. [CrossRef]
18. Ardakanian, O.; Bhattacharya, A.; Culler, D. Non-intrusive occupancy monitoring for energy conservation in commercial buildings. *Energy Build.* **2018**, *179*, 311–323. [CrossRef]
19. Razavi, R.; Gharipour, A.; Fleury, M.; Akpan, I.J. Occupancy detection of residential buildings using smart meter data: A large-scale study. *Energy Build.* **2019**, *183*, 195–208. [CrossRef]
20. Simma, K.C.J.; Mammoli, A.; Bogus, S.M. Real-Time Occupancy Estimation Using WiFi Network to Optimize HVAC Operation. *Procedia Comput. Sci.* **2019**, *155*, 495–502. [CrossRef]
21. Esrafilian-Najafabadi, M.; Haghighat, F. Occupancy-based HVAC control using deep learning algorithms for estimating online preconditioning time in residential buildings. *Energy Build.* **2021**, *252*, 111377. [CrossRef]
22. Deng, Z.; Chen, Q. Reinforcement learning of occupant behavior model for cross-building transfer learning to various HVAC control systems. *Energy Build.* **2021**, *238*, 110860. [CrossRef]
23. Wei, T.; Wang, Y.; Zhu, Q. Deep Reinforcement Learning for Building HVAC Control. In Proceedings of the 54th Annual Design Automation Conference, Austin, TX, USA, 18–22 June 2017. [CrossRef]
24. Escobar, L.M.; Aguilar, J.; Garces-Jimenez, A.; De Mesa, J.A.G.; Gomez-Pulido, J.M. Advanced fuzzy-logic-based context-driven control for HVAC management systems in buildings. *IEEE Access* **2020**, *8*, 16111–16126. [CrossRef]
25. Lou, R.; Hallinan, K.P.; Huang, K.; Reissman, T. Smart Wifi Thermostat-Enabled Thermal Comfort Control in Residences. *Sustainability* **2020**, *12*, 1919. [CrossRef]
26. Wang, C.; Pattawi, K.; Lee, H. Energy saving impact of occupancy-driven thermostat for residential buildings. *Energy Build.* **2020**, *211*, 109791. [CrossRef]
27. Yang, Y.; Hu, G.; Spanos, C.J. HVAC Energy Cost Optimization for a Multizone Building via a Decentralized Approach. *IEEE Trans. Autom. Sci. Eng.* **2020**, *17*, 1950–1960. [CrossRef]
28. Khan, K.H.; Ryan, C.; Abebe, E. Day Ahead Scheduling to Optimize Industrial HVAC Energy Cost Based on Peak/OFF-Peak Tariff and Weather Forecasting. *IEEE Access* **2017**, *5*, 21684–21693. [CrossRef]
29. Khorram, M.; Faria, P.; Abrishambaf, O.; Vale, Z. Air conditioner consumption optimization in an office building considering user comfort. *Energy Rep.* **2020**, *6*, 120–126. [CrossRef]
30. Corbin, C.D.; Makhmalbaf, A.; Huang, S.; Mendon, V.V.; Zhao, M.; Somasundaram, S.; Liu, G.; Ngo, H.; Katipamula, S. *Transactive Control of Commercial Building HVAC Systems*; Pacific Northwest National Lab.(PNNL): Richland, WA, USA, 2016.
31. Mutis, I.; Ambekar, A.; Joshi, V. Real-time space occupancy sensing and human motion analysis using deep learning for indoor air quality control. *Autom. Constr.* **2020**, *116*, 103237. [CrossRef]
32. Yu, L.; Sun, Y.; Xu, Z.; Shen, C.; Yue, D.; Jiang, T.; Guan, X. Multi-Agent Deep Reinforcement Learning for HVAC Control in Commercial Buildings. *IEEE Trans. Smart Grid* **2021**, *12*, 407–419. [CrossRef]
33. Aguilar, J.; Garcès-Jimènez, A.; Gallego-Salvador, N.; de Mesa, J.A.G.; Gomez-Pulido, J.M.; Garcìa-Tejedor, À.J. Autonomic management architecture for multi-HVAC systems in smart buildings. *IEEE Access* **2019**, *7*, 123402–123415. [CrossRef]
34. Garces-Jimenez, A.; Gomez-Pulido, J.-M.; Gallego-Salvador, N.; Garcia-Tejedor, A.J. Genetic and Swarm Algorithms for Optimizing the Control of Building HVAC Systems Using Real Data: A Comparative Study. *Mathematics* **2021**, *9*, 2181. [CrossRef]
35. Collier, M.; Llorens, H.U. Deep Contextual Multi-armed Bandits. *arXiv* **2018**, arXiv:1807.09809.
36. Riquelme, C.; Tucker, G.; Snoek, J. Deep Bayesian bandits showdown: An empirical comparison of Bayesian deep networks for Thompson sampling. In Proceedings of the 6th International Conference on Learning Representations, ICLR 2018-Conference Track Proceedings, Vancouver, BC, Canada, 30 April–3 May 2018.

 *energies*

Article

Application of Artificial Neural Networks for Virtual Energy Assessment

Amir Mortazavigazar [1,2], Nourehan Wahba [1], Paul Newsham [1,3], Maharti Triharta [1], Pufan Zheng [1], Tracy Chen [1] and Behzad Rismanchi [1,*]

[1] Renewable Energy and Energy Efficiency Group, Department of Infrastructure Engineering, Melbourne School of Engineering, The University of Melbourne, Melbourne, VIC 3004, Australia; amortazaviga@student.unimelb.edu.au (A.M.); n.wahba@unimelb.edu.au (N.W.); paul.newsham@student.unimelb.edu.au (P.N.); m.triharta@unimelb.edu.au (M.T.); pufanz@student.unimelb.edu.au (P.Z.); chen.s.tracy@gmail.com (T.C.)
[2] Virginia-Maryland College of Veterinary Medicine, Roanoke, VA 24060, USA
[3] Faculty of Business, UQ Business School, Economics & Law, University of Queensland, Brisbane, QLD 4072, Australia
* Correspondence: brismanchi@unimelb.edu.au

Citation: Mortazavigazar, A.; Wahba, N.; Newsham, P.; Triharta, M.; Zheng, P.; Chen, T.; Rismanchi, B. Application of Artificial Neural Networks for Virtual Energy Assessment. *Energies* **2021**, *14*, 8330. https://doi.org/10.3390/en14248330

Academic Editor: Ana-Belén Gil-González

Received: 17 November 2021
Accepted: 8 December 2021
Published: 10 December 2021

Abstract: A Virtual energy assessment (VEA) refers to the assessment of the energy flow in a building without physical data collection. It has been occasionally conducted before the COVID-19 pandemic to residential and commercial buildings. However, there is no established framework method for conducting this type of energy assessment. The COVID-19 pandemic has catalysed the implementation of remote energy assessments and remote facility management. In this paper, a novel framework for VEA is developed and tested on case study buildings at the University of Melbourne. The proposed method is a hybrid of top-down and bottom-up approaches: gathering the general information of the building and the historical data, in addition to investigating and modelling the electrical consumption with artificial neural network (ANN) with a projection of the future consumption. Through sensitivity analysis, the outdoor temperature was found to be the most sensitive (influential) parameter to electrical consumption. The lockdown of the buildings provided invaluable opportunities to assess electrical baseload with zero occupancies and usage of the building. Furthermore, comparison of the baseload with the consumption projection through ANN modelling accurately quantifies the energy consumption attributed to occupation and operational use, referred to as 'operational energy' in this paper. Differentiation and quantification of the baseload and operational energy may aid in energy conservation measures that specifically target to minimise these two distinct energy consumptions.

Keywords: virtual energy assessment; artificial neural network; commercial buildings; energy efficiency; energy saving

1. Introduction

Energy efficiency is the hidden fuel for meeting the current global demand, yet it remains underutilised despite its proven potentials [1]. According to the 2018 International Energy Efficiency Scorecard Report [2], Australia has ranked 17th out of the top 25 countries consuming 78% of all energy on the planet, indicating its urgency to rally innovative endeavour in the field of energy efficiency. In conjunction with the latter, the state of Victoria's Climate Change Act 2017 targets net-zero greenhouse gas emissions by 2050, which includes the transformation of the commercial building sector to encounter the pressure from a rapidly growing population to building emissions [3]. To reduce the gradual rise of building energy consumption, it is therefore critical to develop a holistic framework in identifying energy flow and waste within a building, replacing the traditional periodic walk-through energy audits. Furthermore, the gradual rise of energy cost has propelled the urgencies of energy conservation and efficiency, which created a need for energy audits to be conducted regularly in commercial and industrial sectors [4,5].

An energy audit is known as the first step to identify energy flow within a building, which helps to achieve energy savings, improve indoor user comfort and productivity, and reduce environmental impact [6]. This assessment can be used by large and small businesses and even households to achieve higher energy efficiency improvements and carbon emissions reductions. A survey from the U.S. Energy Department shows that energy audits have saved about 20% of energy for the commercial building as of 2020 and are expected to reach 30% by 2030 with the utilisation of current technology. However, if emerging technologies are introduced, this number is expected to reach 55% within five years [7]. Joshi [8] claims that a reduction in the amount of energy input can be achieved without reducing the level of useful output services, which leads to the implementation of energy conservation measures throughout a building's lifecycle as an emerging response to the rising energy consumption and carbon emissions of the global building sector. A traditional energy audit involves physical functional observation of the building, conducting a series of data acquisitions of onsite indoor quality parameters, and operational tests such as leak tests, infrared imaging, blower door tests, and equipment sub-metering (as defined by the standards in [6]).

Traditional onsite energy audits require a substantial financial commitment of human resources allocation, travelling times, and installation of instrumentation. In some circumstances, overall implementation and the energy audit cost could exceed the potential energy saving cost. A traditional energy audit provides a periodic snapshot of the building energy behaviour [9]. On the other hand, virtual energy assessment (VEA) does not involve a physical walk-around of the building and aims to cover the activities of the physical energy audit remotely without travelling time and allocation of physical auditor onsite [10]. Conceivably VEA is a way to save time and cost and, more importantly, it has the potential to provide a more frequent assessment of the building energy behaviour to the extent of autonomous continuous assessment with machine learning. The concept of VEA is not entirely novel, it has been occasionally conducted before the COVID-19 pandemic in several cases of residential and commercial buildings as summarised in Table 1. Literature review of building energy. However, it was not widely used as the preference of building owners was leaning towards the traditional energy audit [11]. The unprecedented time of the COVID-19 pandemic has catalysed the application of VEA and provided the opportunities to unlock the market of VEA.

Energy modelling such as VEA can be conducted through two approaches: the bottom-up approach and the top-down approach [12]. The bottom-up approach entails thermal analysis for an individual building, with data related to the building's enclosure, schedules, external weather conditions, internal loads, and potential systems. A white-box model (a bottom-up approach) is commonly used for virtual energy assessments where the building's prediction relies on simplified heat balance calculations, aggregated weather libraries, fixed schedules, and thousands of building's characteristics inputs. Consequently, inaccurate conclusions of energy efficiency can be reached, leading to a gap between the simulated models and actual building interactions. Conversely, a black-box model (also a bottom-up approach) is based on the actual building's energy consumption datasets to provide insight into the existing building performance without extensive buildings' background and information needed. The artificial neural network (ANN) has been an outperforming tool for VEAs [13].

Table 1. Literature review of building energy assessment approaches.

References	Building Classification/Type of Assessment	Methodology	The Direction of the Study
Virtual home energy auditing at scale [14].	Residential/VEA	Regression Model	The virtual assessment was performed based on top-down modelling approaches. The model was based on large publicly available sample data of residential houses from one region and has never been tested in another region. Furthermore, the use of publicly available data might be subjected to incorrect entries and distort the accuracy of the models.
Using artificial neural networks to assess HVAC-related energy saving in retrofitted office buildings [15].	Commercial/VEA	Artificial Neural Network and Multiple Linear Regression (MLR) Model	Two prediction models were developed: MLR and ANN (feedforward multilayer perceptron), using large datasets obtained during energy audits. ANN has superior performance to the MLR model. However, it lacks explanations on its internal parameters and takes longer training time on a trial-and-error basis, MLR model provides a transparent understanding of the linear relationship between the dependent and independent variables. The variable selection process is similar to both models and the variables selected are overlapping. There may be other variables that have been not considered in the process.
Neural networks for smart homes and energy efficiency [16].	Residential/N.A.	Neural Network	The paper discussed theoretical approaches of self-regulated heating system of each unit in a communal housing by a smart home system which include neural networks that were trained in the tenant preferences using acquired data from sensors and live feedback. A simple recurrent network was deemed sufficiently effective however the appropriate function depends on the required number of dimensions and output data. The discussion did not include any examples where the approach was practically implemented.
Energy analysis of a building using artificial neural network: A review [17].	Various Building classification/N.A.	Neural Network	The paper reviewed diverse applications of ANN in the prediction of building energy consumption, with the three most used networks being feedforward, competitive, and recurrent networks. The paper also stated that indoor air temperature is often regarded as the only control variable whilst another thermal comfort factor such as humidity was rarely considered, hence it might be beneficial to develop control strategies based on thermal comfort. Performance and adaptability for a constantly changing environment of ANN models needed to be considered as well.
Energy audits in industrial processes [18].	Industrial and Commercial/Traditional onsite	Various auditing tools such as Heating Assessment and Survey Tool programming	Energy efficiency measures were gauged for six industrial processes case studies to reduce the fuel consumption in the U.S. The procedure followed for energy assessing targeted specific processes and depended on walk-through bottom-up approaches and basic thermal analysis tools. The dependency on averaging and simple calculations in the case studies had led to overestimating the energy consumption reduction. In addition, the wide variety of energy processes limits the versatility of auditing procedures, which should only describe a broad framework of audits.
Application of multiple linear regression and an artificial neural network model for the heating performance analysis [19].	Commercial/N.A.	Artificial Neural Network and Regression Model	MLR and ANN models were developed for the measurement and verification baseline for probable future energy conservation measures in a ground source heat pump system (GSHP). Various MLR models were developed to specify the influencing factors in the GSHP performance and establish prediction accuracy for the optimal ANN architecture. The deep belief network (DBN) was used as the ANN model, to counter the impact of backpropagation sensitivity. This research highlighted the potential future application of ANN as a smart energy audit tool to provide energy conservation solutions.
Applying computer-based simulation to energy auditing [20].	Commercial/N.A.	eQuest simulation software tool	A bottom-up approach has been investigated through a case study of a high-rise tower in the U.S. The energy assessment required extensive knowledge of the building architecture and calibration, in addition to the building internal loads and HVAC systems. The research pinpointed the limitations imposed by data such as information accessibility which prohibit the models from reflecting the reality.
Random Forest-based hourly building energy prediction [21]	Commercial (Educational)/NA	Random Forest prediction model	This paper proposed the use of a random forest prediction model to estimate the hourly energy consumption of a building. Randomisation of building variables is applied to generate initial training sets to develop a tree splitting process based on a collection of regression trees. The performance of the random forest prediction model was tested on educational buildings at the University of Florida. The paper showcased the ability of the random forest algorithm to predict hourly energy consumption.

Table 1 implies that VEA and neural networks are commonly regarded as two independent concepts. It presents various research of building energy assessment approaches that paved the boost of machine learning tools in Table 1 uncommonly used VEA. With the vast accessibility to huge, recorded databases, the rapid growth of machine learning (ML) tools utilisation such as ANN is facilitated to extract valuable insights and predictions to assist energy consumption performance [22]. The ANN applications are surging as one of the most popular artificial intelligence (A.I.) models. Data analytics, supported by machine learning models and big data, has the potential to explore new solutions for pressing energy consumption issues [23].

In line with Victoria' Climate Change Act 2017 and Australia's national target, the paper proposes a framework with a hybrid of top-down and bottom-up approaches for a VEA, implementing ANN and conducting a "virtual" walk-around simultaneously. This framework is demonstrated using case study buildings on the University of Melbourne campus to highlight the potential of the energy gap during COVID-19 lockdown. Hence, this application can contribute to the University of Melbourne Sustainability plan 2022–2025 [24] to achieve Victoria's commitment to reduce greenhouse gas emissions from electricity consumption in 2030.

The methodology is described in Section 2, which is parted into three sub-sections of data collection and quality checks, modelling with neural networks, and uncertainty analysis. The results are presented in Section 3, which depict the historical timeline of the electrical consumption, the performance of neural networks with a confidence interval, and occupancy correlation to the electrical consumption. Discussions of building base load, parameters sensitivities, and future works of the VEA are presented in Section 4 Discussion.

2. Methodology

In this paper, a hybrid methodology was developed consisting of exploratory data collection for the cases study buildings and computer-aided neural network modelling using MATLAB, thus depicting the hybrid approach of top-down and bottom-up approaches. This section started with the description of the buildings being assessed, energy consumption data collection and quality checks, and data modelling with ANN function selection and architecture.

2.1. Case Study

The VEA was performed in four educational buildings at the University of Melbourne Parkville Campus during the COVID-19 pandemic lockdown in August 2020. To showcase the capabilities of the proposed methodology, multifunctional buildings were selected that contained public areas, lecture halls, laboratory facilities and office space. The selected buildings are described in Table 2. The buildings varied in age from 12 to 62 years, usable area of 3000 to 13,000 m^2, and a range of different enclosure materials such as concrete, brick and unglazed/glazed glass.

Table 2. Comparative summary of buildings selected for case study.

Building ID	Building Name	Built (Age)	Usable Area (m^2)	Number Occupied Floors	Building Façade Materials
B1	Alan Gilbert	2001 (20 yr)	9010	12	Concrete, glass windows with glazing
B2	The Spot	2009 (12 yr)	13,140	15	Glass-covered in 50% frit
B3	Baillieu Library	1959 (62 yr)	12,540	8	Steel framed glass, brick
B4	Elec Engineering	1973 (48 yr)	3670	6	Concrete, brick and unglazed windows

2.2. Virtual Energy Assessment

2.2.1. Data Collection and Quality Checks

The case study buildings are equipped with a real-time energy monitoring platform that is accessible through a secure web server. These energy data were collected for the case study buildings for this study, as well as external weather conditions and occupancy rates. The external weather data were collected from the nearest weather station and the occupancy rate was collected from the university security. All of the above-mentioned data could be accessed for most of the new buildings and most of the existing buildings which have been upgraded with remote data access. The authors believe that the proposed methodology is replicable to any building with such datasets. Abnormalities and outliers were removed as part of the data quality checks.

Electrical consumption data was collected for a total load of each building separately using Clariti. End-uses submetering was unavailable, with meter data representing the total load consumption. Fifteen minutes of load consumption was gathered from June 2015 to January 2021, as the basis of the assessment. With the building closures due to COVID-19, the use of space and buildings' layouts and building enclosure details were obtained from facility managers. Furthermore, the electrical consumption data were screened for outliers, the Clariti tool is very precise and for the cases, at hand, no outlier was detected.

Central to the building external conditions, 10 years of detailed meteorological data from the Australian Bureau of Meteorology in 15-min increments were procured from the nearest weather station. This data included temperature, humidity, rainfall, and solar irradiance. Online images of the sides and tops of the buildings were supplemented by using tools such as Google Earth and Nearmap.

2.2.2. Variables Selection

To improve the learning efficiency and compilation of the ANN model, it was necessary to reduce the number of input data variables used to those that were most influential of the energy use in each building. For instance, occupants' behaviour is a typical key element of a building's energy performance. In this paper, occupants' behaviour information was scarce and limited to certain years, influencing its validity in the training process of the neural network; however, if available, it would have provided further fertile testing grounds for the neural network's ability to further detect the end-use energy performance.

2.2.3. Artificial Neural Network Modelling

The advent of ANN enables analysis from the complicated and large size of data, extracting patterns and trends to provide future projections with the potential to be trained for projections of unpredicted circumstances. It is currently widely used with applications ranging from sales forecasting, web searching, to visual imaging. ANN model is a simplified version of a biological neural network that combines data and stores relationships between independent and dependent variables. The ANN can self-study the historical data and users' preferences, through training, to improve its accuracy of prediction. It can also continue learning, which can be well adapted to a new environment [25]. Compared to sophisticated calculations of statistical models, the ANN model trains data for prediction, which is more suitable for a larger set of data [17].

As illustrated in Figure 1, a basic neural network consists of several connected nodes, or "neurons", which produce a sequence of activation. Input neurons (in the input layer) are activated using an initialiser, while other neurons in hidden layers are activated through weighted connections from previously activated neurons. After checking the outcome with desired outputs, the neurons adjust their weight functions to get a more accurate result [17]. This procedure is technically called "learning". In general, the traditional-learning approach in a neural network requires significant amounts of data and long chains of computational stages to obtain accurate results. To handle this obstacle, various other networks are developed to accurately assign the weight functions by adopting the training stages with an unsupervised learning technique.

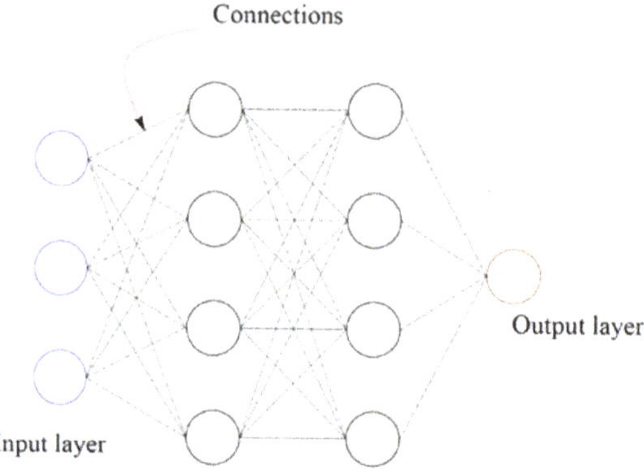

Figure 1. A typical Feedforward neural network with three input variables, one output variable and 4 × 2 hidden layers.

A branch of the ANN model, the Nonlinear Autoregressive Network with Exogenous Inputs (NARX), is a recurrent dynamic network with feedback connection enclosing several layers of the network to account for time-series modelling [26]. A NARX model has less sensitivity to the problem of long-term dependencies and has an outstanding learning capability and generalisation performance for time-series data. The superiority of NARX is also reflected in its ability to model multi-dimensional data and its ability to predict price fluctuations accurately. Compared to logarithmic multiple linear regression and multiple linear regression, NARX artificial neural network has higher accuracy [27].

Figure 2 shows the architecture of the NARX network used in this paper. The main features of this network are 1. number of neurons: $r = 10$; 2. number of input variables: $n = 5$; 3. time lag (delay): $\Delta t = 3$; and 4. number of layers: $N = 1$. The construction of the NARX network encompasses a feedforward network baseline that incorporates the nonlinear regression function of y shown in Equation (1). Where β_{out} is output bias, ω_{out} is output weight, $\varnothing$ is a linear activation function $\varnothing(j) = j$ and, h_k is hidden layer output.

$$y_t = \varnothing \left(\beta_{out} + \sum_{k=1}^{r} \omega_{out\ i} \times h_{kt} \right) \tag{1}$$

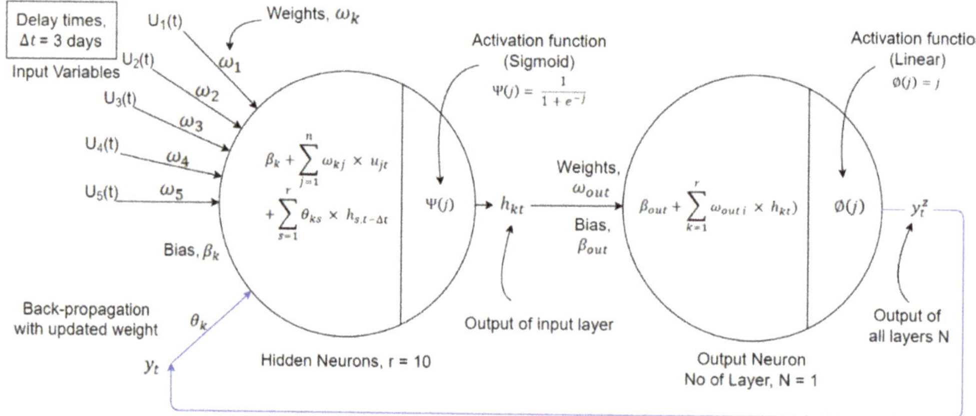

Figure 2. NARX Network's design used in this article (can be replicated) with 5 input variables, 1 output variable, 10 hidden neurons and a delay time of three days.

Equation (2) is the output of the input layer h_k output where β_k input bias, ω_k input weight for input variables, θ_k input weight for the delayed output, Ψ is the sigmoid activation function $\Psi(j) = \frac{1}{1+e^{-j}}$, and h_s is the delayed dynamic autoregression output.

$$h_{kt} = \Psi \left(\beta_k + \sum_{j=1}^{n} \omega_{kj} \times u_{jt} + \sum_{s=1}^{r} \theta_{ks} \times h_{s,t-\Delta t} \right) \tag{2}$$

Now if Equation (2) is inserted in Equation (1), Equation (3) can be drawn. Equation (3) demonstrated the mathematical operations within one NARX layer.

$$y_t = \varnothing \left[\beta_{out} + \sum_{k=1}^{r} \omega_{out\,i} \times \Psi \left(\beta_k + \sum_{j=1}^{n} \omega_{kj} \times u_{jt} + \sum_{s=1}^{r} \theta_{ks} \times h_{s,t-\Delta t} \right) \right] \tag{3}$$

To incorporate all layers of NARX into one equation, Equation (3) can be reproduced as Equation (4) where the operations of all layers are demonstrated.

$$y_t^z = \varnothing \left[\beta_{out}^z + \sum_{z=1}^{N} \sum_{k=1}^{r} \omega_{out\,i}^z \times \Psi \left(\beta_k^z + \sum_{j=1}^{n} \omega_{kj}^z \times u_{jt} + \sum_{s=1}^{r} \theta_{ks} \times h_{s,t-\Delta t} \right) \right] \tag{4}$$

2.2.4. Neural Network Training

Three neural networks were trained to find the optimal model architecture. These were: Multilayer Perceptron, Feedforward, and NARX. The following methodology is tailored to the NARX architecture. The NARX model was fed a 70/30 split of data (70% training, 15% validation, and 15% testing).

Figure 3 demonstrates the input-output data and the process of training the NARX network. As shown in Figure 3 input data consists of five independent variables that were recorded during the period of interest (i.e., before COVID-19 (March 2015–March 2020)). The output variable is the electricity consumption for the building of interest also recorded for the period of interest. The frequency for recording the input and output variables is 15-min intervals. It should be noted that although timesteps are used for the indexing of all variables time is not an input variable.

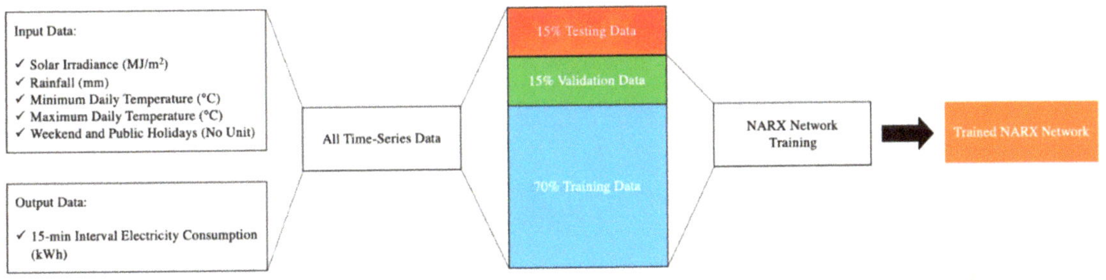

Figure 3. NARX Network training phase using training and validation data (85% of all data).

2.2.5. Neural Network Validation

Now to verify the accuracy of the network and test for overfitting, the trained network is validated using the remaining 15% data using the process shown in Figure 4.

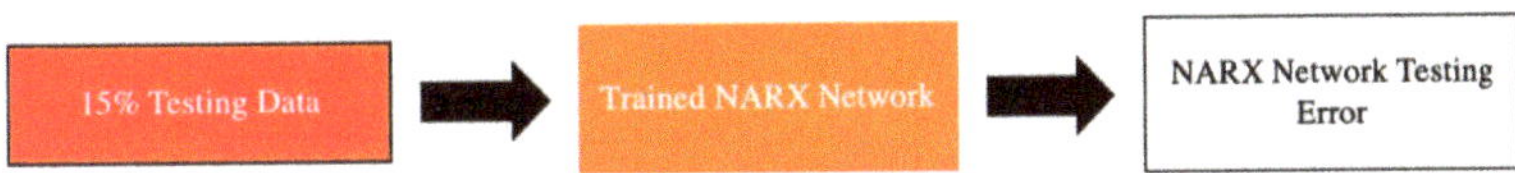

Figure 4. NARX Network testing phase using testing data (15% of all data).

Comparing the 3 neural network architectures below, the NARX network gave the most accurate prediction (Table 3) without systematic biases. Annual cyclical variation in electricity use is also evident.

Table 3. Feedforward, CNN and NARX networks mean absolute percentage error for all buildings.

Building ID	Feedforward MAPE (%)	CNN MAPE (%)	NARX MAPE (%)
B1	19	19	6
B2	14	14	7
B3	9	9	3
B4	15	16	7

The specifics of each network that was used in Table 3 are as follows: (1) Feedforward: This network is a simple Feedforward network using a backpropagation training function and 10 hidden layers that can do some accurate predictions; (2) CNN: This is a conventional neural network but, this article, uses a backpropagation training function and 10 hidden layers; (3) NARX: This network is extensively discussed in the previous section. Figure 2 shows that this network uses 10 hidden layers and also a backpropagation module that uses updated weights.

Table 3 also demonstrates that the NARX network is demonstrating a better MAPE compared to the other networks. This could be referred to as the inherent elements that are embedded into the NARX network (Figure 2) that can be called an improvement to the CNN and Feedforward networks. NARX network not only uses the backpropagation method but also uses delayed inputs and modified weighting that will improve the outputs. This evolution of networks from Feedforward to CNN (with backpropagation) and finally to NARX further demonstrates how critical is the network design considerations.

Figure 5 shows the individual R values for the training, testing and all data inputs for the NARX training and testing process. Overfitting occurs when the network is overtrained on the training dataset and can only produce accurate results for the trained data and less accurate results for the test data set. The results presented in Figure 5 demonstrate that the issue of overfitting has not occurred as the testing data shows a highly accurate result.

Figure 5. *Cont.*

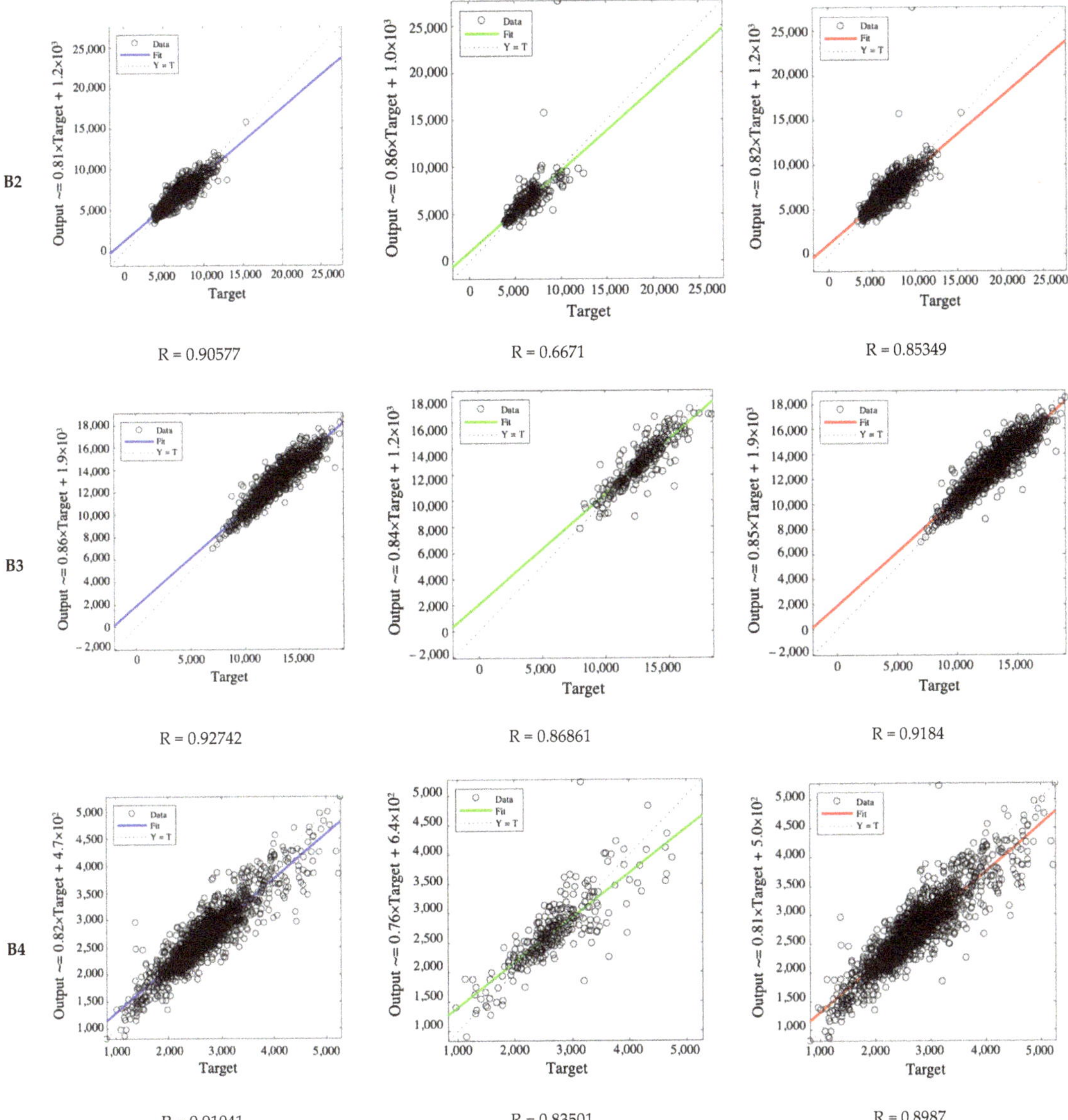

R = 0.90577

R = 0.6671

R = 0.85349

R = 0.92742

R = 0.86861

R = 0.9184

R = 0.91041

R = 0.83501

R = 0.8987

Figure 5. The NARX network outputs for the training, testing, and all datasets.

After completing the network training with the training dataset and then testing with the testing dataset, the network is used to produce a complete prediction using existing datasets. Figure 6 shows the network prediction vs. actual electricity load for all buildings in this paper. Figure 6 shows that the NARX network prediction accurately maps the actual electricity load and fluctuations.

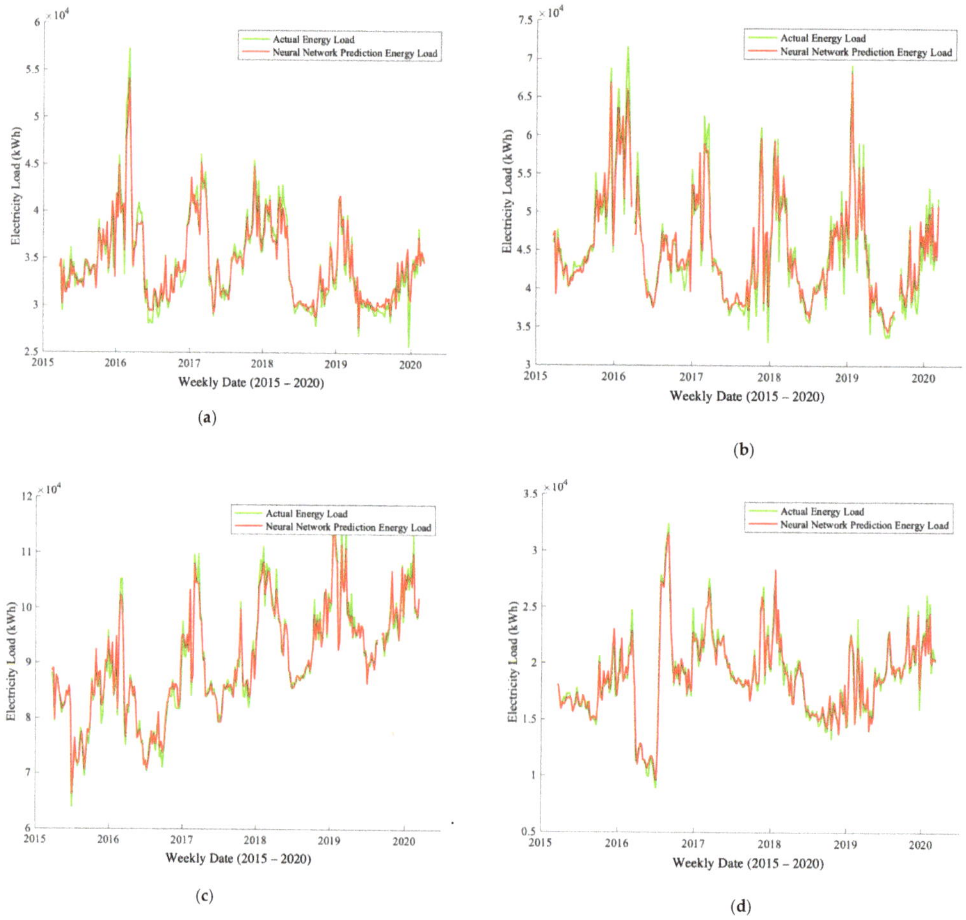

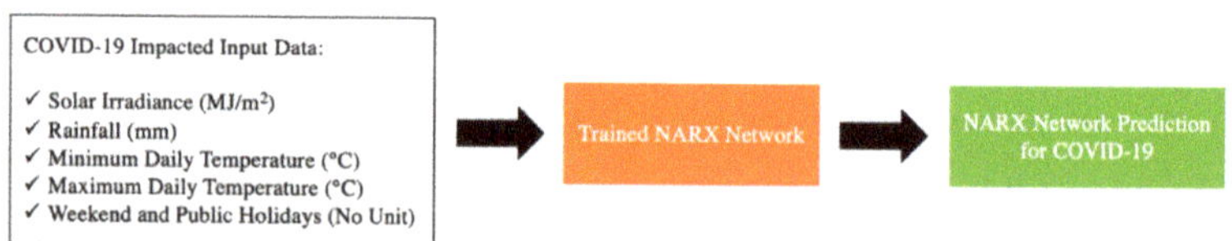

Figure 6. The NARX network prediction vs. actual electricity load for all buildings. (**a**) B1, (**b**) B2, (**c**) B3, and (**d**) B4.

2.3. Neural Network Forecasting

With the NARX network validated and trained following the Figure 7 process, the COVID-19 impacted input data was then given to the network and the network prediction was compared with the actual electricity load. These comparisons will be presented in the next section.

Figure 7. NARX Network prediction phase using COVID-19 impacted period.

3. Network Results Evaluation

3.1. Baseline Due to Operation Interruption Caused by COVID-19

Figures 2–5 below compare the predicted electricity consumption from the NARX model and the actual metered energy consumption for each of the case study buildings.

They include one year of regular operation leading to March 2020 for comparison of the COVID-19 impacted period from March 2020 to January 2021. The metered energy use data from March 2020 to January 2021 show the reduction in electricity use in the buildings during the period, and provide a better understanding of the base electrical load of the buildings over a significant length of time whilst the city of Melbourne endured two significant lockdowns due to health orders from the local spread of the COVID-19 coronavirus:

- First lockdown: 17 March to 1 June 2020. Restriction "easing" began in Victoria accessed on 29 October 2021 (https://www.dhhs.vic.gov.au/coronavirus-update-victoria-1-june-2020), however, new outbreaks caused restrictions to reverse, and started tightening again on 22 June accessed on 29 October 2021 (https://www.dhhs.vic.gov.au/coronavirus-update-victoria-22-june-2020)
- Second lockdown: 8 July to 8 November 2020 accessed on 29 October 2021 (https://en.wikipedia.org/wiki/COVID-19_pandemic_in_Australia#Melbourne)
- Summer period: 30 November 2020, to 28 February 2021, accessed on 29 October 2021 (https://www.unimelb.edu.au/dates/semester-dates)
- Staff and graduate students were beginning to return in late 2020, however, the University did not re-open to students until 3 January 2021.

The impacts of these events were notable in the metered energy use (green) from 17 March 2020, onward in the figures below, though not consistent for each building depending on what activity remained throughout the lockdowns. For B1, B2, and B4 (Figure 8 a, b and d respectively), between March 2020 and January 2021 the lower bound of the actual data is a strong measurement of the building baseload on evenings and weekends, and the upper bound was a strong indicator of the daytime building loads without added thermal load due to occupancy.

The situation was different for B3 (Library) (Figure 8c), which remained open with minimal staff to provide continued support to the university staff and students continuing working and studying remotely during the study period. Electricity consumption was lower than predicted and irregular, but overall, due to the continued activity, there was no definitive measurement of the baseload captured for the library. During the break between semesters in July 2020, the library had higher than predicted electrical consumption, which aligned with the short periods between the first and second lockdowns and the start of the second delayed semester of the 2020 school year, possibly due to the significant activity surrounding planning for a full semester to be completed online due to the lockdown.

Some university services beyond libraries were open during the lockdowns, identified as essential services, albeit only for staff to be onsite for student support and filling student requests for pickup. Similarly, some faculties were also permitted to continue where online studies were not possible (e.g., medical, biological labs, etc.), but under strict health protocols as directed by the government of Victoria.

Overall, the amplitude of the metered energy use was decreased significantly between daytime peaks and baseload. The spread between the actual overnight vs. day baseload is roughly 20% of the magnitude spread of the predicted electrical consumption (red), where the remaining 80% avoided electrical consumption could be attributed to the occupation/operational use of the building. The offset between the lower bounds of the predicted and actual could be due to the additional thermal mass accumulation and additional HVAC programs during regular operation (with added plug loads).

Limited information was available from the university facilities departments for what changes were made to the building's internal systems for energy savings (if at all) when the lockdown was initially announced in March 2020. Assumptions were made that B1, B2, and B4 had their HVAC systems put to holiday settings, and some IT systems such as the A/V in the lecture halls were powered down. In September 2020, leading into the hotter summer weather in Melbourne, additional measures were taken to reduce HVAC operation by shutting curtains and minimising lighting demand to only safety/security requirements.

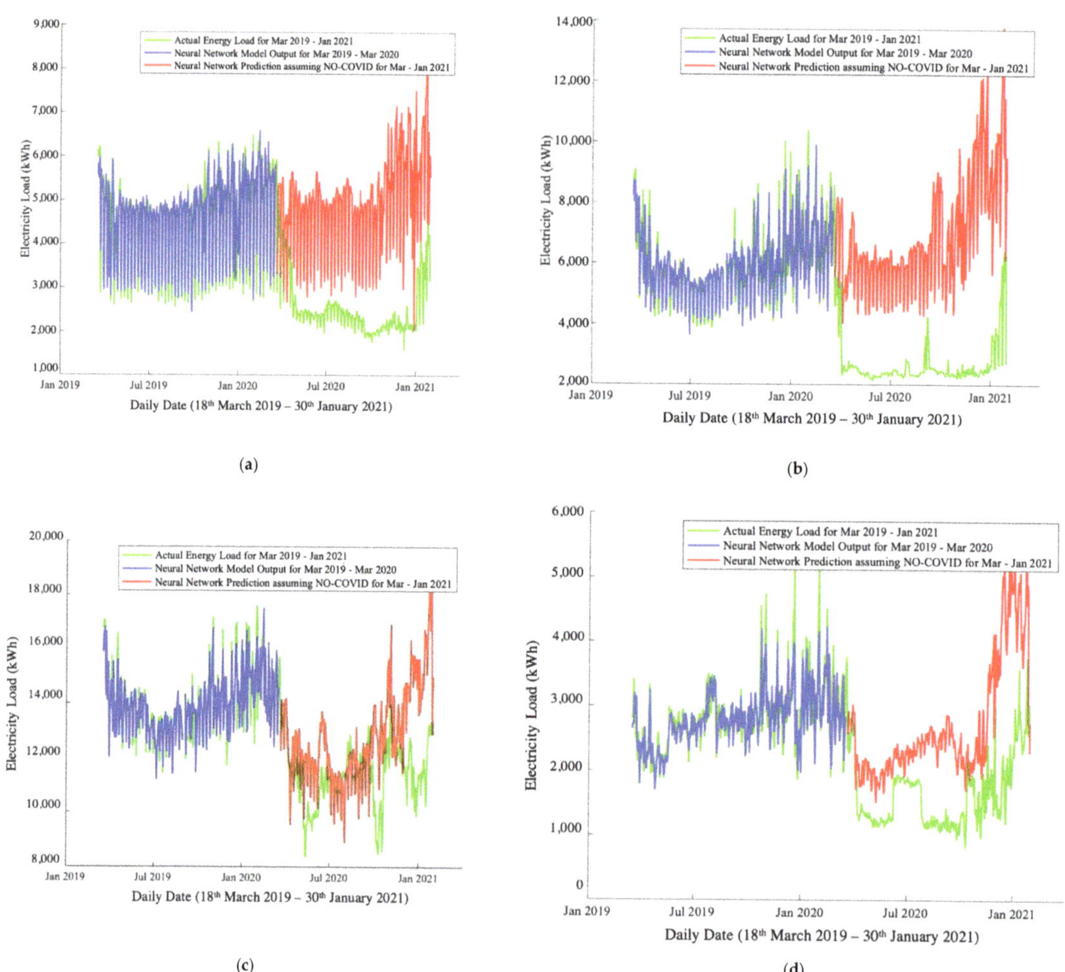

Figure 8. NARX Prediction for COVID-19 impact for (**a**) B1, (**b**) B2, (**c**) B3, and (**d**) B4.

3.2. Sensitivity Analysis

A sensitivity analysis was applied to determine the impact of key variable conditions, contributing to the building's energy consumption forecast and the VEA accuracy. To understand the influence of each of these conditions input on the neural network output, the meta-model-based sensitivity method of white gaussian noise was implemented to case studies data sets, to randomly generate noise values and retrain the NARX network. This is a sampling-based probabilistic method to maintain a well-validated model code. Once the retraining process was over, the mean absolute percentage error of each of the regenerated data sets was calculated and compared with the performance of other neural networks' architectures, leading us to determine the most influential data sets impacting the performance of the neural networks and the forecast certainty [28]. From Figure 9, the most influential variable of the neural model performance for B1, B2, and B3 was maximum temperature, while the most influential factors for the electrical engineering building were the maximum temperature and solar irradiance. This was due to the variation of the buildings' envelope material from Table 2, as B4 fenestration material was unglazed glass.

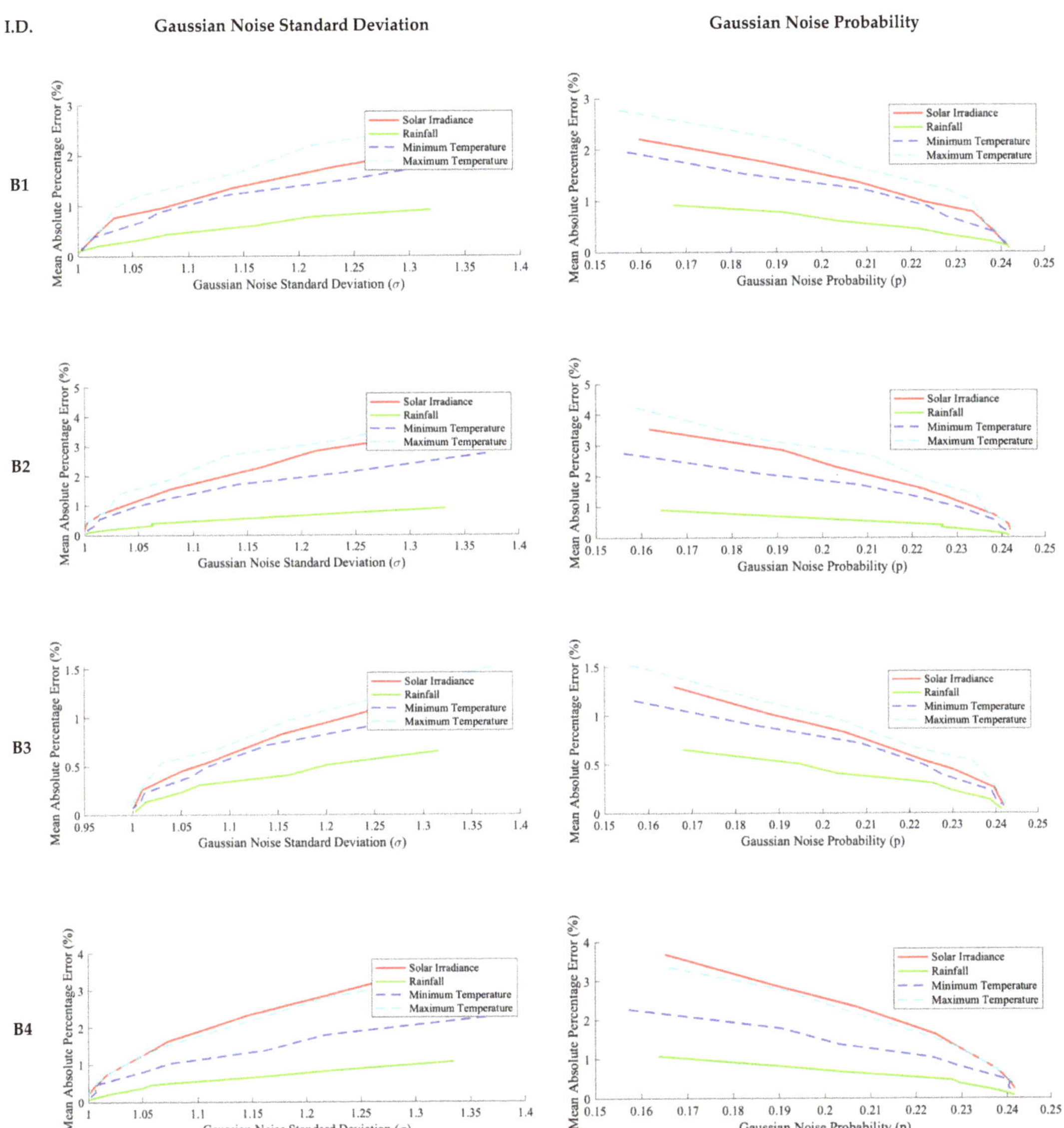

Figure 9. The NARX network sensitivity analysis for four input variables: solar irradiance, rainfall, and minimum temp and maximum temp.

3.3. Uncertainty Analysis

Although NARX networks have been employed for several time series applications, they are not immune to uncertainty caused by several factors such as vanishing and exploding gradients, inappropriate selection of the neural network architecture, and false convergence to local optima instead of global optima during the training process [29]. To quantify the level of uncertainty associated with the energy consumption forecast, uncertainty analysis was conducted (Figure 10) by using confidence interval principles for unknown sample distribution. Energy consumption of the case study buildings is based on stochastic weather data, building use, occupancy rate, and unplanned events such as COVID-19 lockdown. Expressing the forecast energy consumption results with prediction intervals provide certainty to the neural network outputs. The predicted energy consumption population sample has an unknown mean $\overline{x}$ and unknown standard deviation $\frac{s}{\sqrt{n}}$. The concept of unknown sample distribution states that the sample standard deviation is equivalent to an estimated standard error, replacing the standard deviation values, where the standard error approaches an equal value of the standard deviation for a large sample number n [30]. The confidence interval for the sample corresponds to the 95% (i.e., 1–2 α level) upper and lower levels of confidence, representing the 2.5th (q_L) and 97.5th (q_L) percentiles of the distribution of every simulated monthly prediction. This can be created as (5).

$$\overline{x} \pm t^* \frac{s}{\sqrt{n}} \tag{5}$$

where $\overline{x}$ is the monthly average of the predicted energy consumption, t^* is the critical value of the t distribution with unlimited degrees of freedom, and $\frac{s}{\sqrt{n}}$ is the standard error of the monthly predicted values.

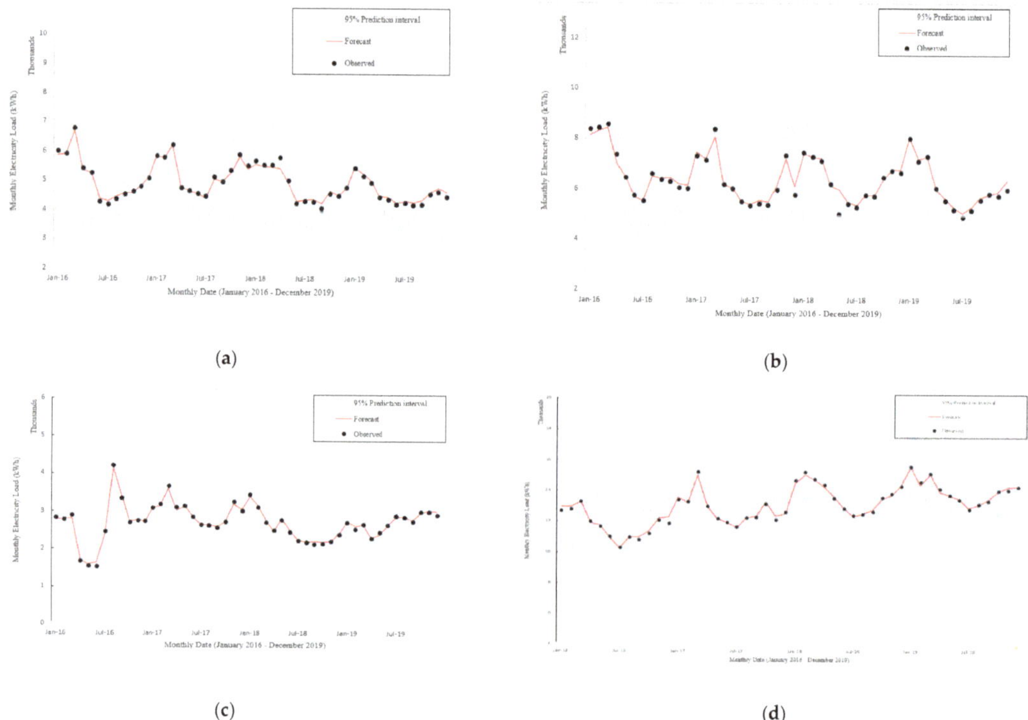

Figure 10. The 95% prediction interval for NARX forecast vs. observed electricity load for all buildings. (**a**) B1, (**b**) B2, (**c**) B3, and (**d**) B4.

To measure and compare the uncertainty based on each case study and model, p-factor and d-factor are used as the objective criteria [31]. The "P-factor" indicates the percentage of observed data as the actual energy consumption bracketed by the confidence interval upper and lower limit. The D-factor is the average distance between the upper and lower limits, calculated as follows:

$$\overline{d_x} = \frac{1}{n} \sum_{i=1}^{n} (X_u - X_l) \tag{6}$$

$$d - factor = \frac{\overline{d_x}}{\sigma_x} \tag{7}$$

where $\overline{d_x}$ is the average distance between the upper X_u and lower levels X_l of the confidence band for each month, and σ_x is the standard deviation of the observed energy consumption [32]. The best outcome of the uncertainty analysis of a simulated prediction is to have 100% of the P-factor falling within the confidence band, and a D-factor close to zero. Table 4 indicates that the observations of the electricity consumption for the chosen buildings fall entirely within the prediction interval of the forecast energy consumption. Thus, this analysis eliminates the uncertainty associated with the NARX energy consumption predictions.

Table 4. The NARX uncertainty analysis P-factor and D-factor.

Building ID	P-Factor (%)	D-Factor
B1	100%	0.16
B2	100%	0.09
B3	100%	0.08
B4	100%	0.09

4. Discussion

The unprecedented time of the COVID-19 pandemic presented a unique situation that has unlocked the market of VEA and boosted its application commercially. The authors developed and validated a numerical model for VEA and have demonstrated the tool through selected multifunctional case study buildings. The advantages and limitations of the model are discussed in this section.

One of the advantages of the model is evaluating the energy gap during unprecedented times. From Figures 3, 4 and 6, the NARX model provides a reasonable forecast horizon for the typical energy usage from March 2020 to January 2021. This has provided a rare opportunity to accurately quantify the energy use attributed to occupation and operational use of the building (i.e., the difference between typical building energy use vs. unoccupied building base load during COVID-19, defined as the "energy gap"). This energy gap shifts more emphasis on behavioural energy efficiency measures to reduce the operational energy load.

Central to the framework approach proposed in this paper, a hybrid of bottom-up and top-down approaches was developed, using a 'black-box' model with feature extraction, respectively.

A black-box model, such as the NARX neural network employed in this study, enabled a virtual study of a building envelope with an unknown internal system, with external input data such as weather and solar exposure generating the outputs of the building's energy use. This was beneficial to gain insights into the building operation with minimal input data, which would have been difficult with common bottom-up white-box methods, which require substantial upfront knowledge of the building's construction [12]. In addition, the neural network was able to cope with non-linearity and the stochastic nature of the raw input temperature and raw output energy data, which was not considered with the white box methods that use averaged and typical data files for more generalised results. This inferred that the NARX should provide significantly more reliable results in prediction, especially if a significant amount of historical data is available.

However, a notable limitation of the black-box approach would be the limited adaptability of the model to other buildings. A neural network is tailored to the building on which the model was trained upon, which prevents the extrapolation of results for use to other buildings and therefore results in the limited macro-level understanding of the campus. Predictions were only as good as the input data, and hence the NARX model can only produce predictions as accurate as of the input data for weather and energy use. Deb and Schlueter [12] similarly noted the top-down approaches were "unsuitable for individual building retrofits", hindering the proportioning of the total consumption into energy end uses.

The NARX model was supplemented by the top-down 'feature extraction method' represented in the sensitivity analysis to determine the influential factors causing the variations of energy consumption based on the case study buildings' ages and envelope material. The sensitivity analysis prioritised the input data variables required in the description of the building energy consumption, as well as reducing the training time of the NARX model. In this study, the most influential variables to the buildings' energy consumption were hourly meteorological maximum temperature and solar irradiance, as shown earlier in Figure 9.

This combination of the two approaches has enabled the VEA to avoid the uncertainty inherited from the bottom-up aggregated simplified approaches and to gain an understanding of the interactions of the building beyond the black box limitations.

Further work could be conducted to examine the resiliency of each building, by expanding the envelope of these influential variables to account for more extreme climate scenarios. This could provide valuable insight into how a building would react in its current form in such scenarios and guide future retrofit priorities and design more tailored and sustainable energy conservation measures.

5. Conclusions

The unprecedented time of the COVID-19 pandemic presented a unique situation that has unlocked the market of VEA and boosted its application commercially. The authors developed and validated a numerical model for VEA and have demonstrated the tool through a case study of multifunctional educational buildings. The advantages and limitations of the model have been elaborated in detail. The key opportunity presented was to evaluate the "operational energy" over a significant amount of time which encompassed all four seasons in continuity (i.e., the difference between typical building energy use vs. unoccupied building base load during COVID-19 as demonstrated in [33]). In this paper, a combination of bottom-up ('black-box' model) and top-down ('feature extraction') approaches were employed to conduct the VEA. This combination reduced the inherent uncertainty from typical bottom-up aggregated simplified approaches under similar data collection scenarios and provides meaningful insight into the behaviour of the building beyond employing only the single 'black-box' methodology. A neural network model has limited applicability since it is tailored using the data attributed to that specific building. Supplementing the NARX model with a separate top-down approach (i.e., the sensitivity analysis) identified the most influential factors defining the variations of energy consumption, which not only optimised model performance but provided insight for where to target ECM activity for each building.

Author Contributions: Conceptualisation; B.R., A.M., P.N. and N.W., methodology; A.M., N.W., M.T. and P.N., software; A.M. and N.W., validation; B.R., A.M. and N.W., formal analysis; M.T., N.W. and P.N., investigation; A.M. and P.N., resources; A.M. and P.N., data curation; A.M., P.N. and P.Z., writing—original draft; A.M., P.N., N.W., M.T., P.Z. and T.C., writing—review and editing; B.R. and N.W., visualization; A.M., M.T., N.W. and T.C., supervision; B.R. All authors have read and agreed to the published version of the manuscript.

Funding: This research received no external funding.

Institutional Review Board Statement: Not applicable.

Informed Consent Statement: Not applicable.

Data Availability Statement: Data used in this article are publicly available through the University of Melbourne Sustainable Campus initiative using Clariti. https://sustainablecampus.unimelb.edu.au, accessed on 1 May 2021.

Acknowledgments: This work was conducted within the Department of Infrastructure Engineering as a part of the ongoing energy assessment of the university campus. The authors would like to thank the University of Melbourne Campus Services for providing raw data for energy consumption and occupancy density.

Conflicts of Interest: The authors declare no conflict of interest.

References

1. IEA. *Net Zero by 2050*; IEA: Paris, France, 2021. Available online: https://iea.blob.core.windows.net/assets/deebef5d-0c34-4539-9d0c-10b13d840027/NetZeroby2050-ARoadmapfortheGlobalEnergySector_CORR.pdf (accessed on 15 October 2021).
2. Castro-Alvarez, F.; Vaidyanathan, S.; Bastian, H.; King, J. *The 2018 International Energy Efficiency Scorecard*; American Council for an Energy-Efficient Economy: Washington, DC, USA, 2018.
3. Victoria, T.P.O. *Climate Change Act 2017*; Legislation Victoria, 2017. Available online: https://www.climatechange.vic.gov.au/legislation/climate-change-act-2017 (accessed on 15 October 2021).
4. Cronin, J.; Anandarajah, G.; Dessens, O. Climate change impacts on the energy system: A review of trends and gaps. *Clim. Chang.* **2018**, *151*, 79–93. [CrossRef] [PubMed]
5. Akhmat, G.; Zaman, K.; Shukui, T.; Sajjad, F. Does energy consumption contribute to climate change? Evidence from major regions of the world. *Renew. Sustain. Energy Rev.* **2014**, *36*, 123–134. [CrossRef]
6. Australia, S. *Energy Audits Commercial Buildings AS/NZS 3598.1:2014*; Standards Australia: Sydney, Australia, 2014.
7. Austin, D. *Addressing Market Barriers to Energy Efficiency in Buildings*; Working Papers 43476; Congressional Budget Office: Washington, DC, USA, 2012. Available online: https://ideas.repec.org/p/cbo/wpaper/43476.html (accessed on 15 October 2021).
8. Joshi, A.; Mundada, A.; Suryavanshi, Y.; Kurulekar, M.; Ranade, M.; Jadhav, S.; Patil, K.; Despande, Y. Performance Assessment of Building by Virtual Energy Audit. In Proceedings of the 2018 International Conference and Utility Exhibition on Green Energy for Sustainable Development (ICUE), Phuket, Thailand, 24–26 October 2018.
9. Glick, M.B.; Peppard, E.; Meguro, W. Analysis of Methodology for Scaling up Building Retrofits: Is There a Role for Virtual Energy Audits?—A First Step in Hawai'i, USA. *Energies* **2021**, *14*, 5914. [CrossRef]
10. Cowan, J.; Pearson, R.; Sud, I. *Procedures for Commercial Building Energy Audits*; American Society of Heating; Ashrae: New York, NY, USA, 2004; ISBN 9781936504091. Available online: https://www.techstreet.com/standards/procedures-for-commercial-building-energy-audits-2nd-edition?product_id=1809206#product (accessed on 15 October 2021).
11. Avina, J.M.; Rottmayer, S.P. Virtual Audits: The Promise and The Reality. *Energy Eng.* **2016**, *113*, 34–52. [CrossRef]
12. Deb, C.; Schlueter, A. Review of data-driven energy modelling techniques for building retrofit. *Renew. Sustain. Energy Rev.* **2021**, *144*, 110990. [CrossRef]
13. Li, A.; Xiao, F.; Fan, C.; Hu, M. Development of an ANN-based building energy model for information-poor buildings using transfer learning. *Build. Simul.* **2021**, *14*, 89–101. [CrossRef]
14. Hoşgör, E.; Fischbeck, P.S. Virtual home energy auditing at scale: Predicting residential energy efficiency using publicly available data. *Energy Build.* **2015**, *92*, 67–80. [CrossRef]
15. Deb, C.; Lee, S.E.; Santamouris, M. Using artificial neural networks to assess HVAC related energy saving in retrofitted office buildings. *Sol. Energy* **2018**, *163*, 32–44. [CrossRef]
16. Roessler, F.; Teich, T.; Franke, S. Neural Networks for Smart Homes and Energy Efficiency. In *DAAAM International Scientific Book*; DAAAM International Publishing: Vienna, Austria, 2012; Chapter 26; pp. 305–314. [CrossRef]
17. Kumar, R.; Aggarwal, R.K.; Sharma, J.D. Energy analysis of a building using artificial neural network: A review. *Energy Build.* **2013**, *65*, 352–358. [CrossRef]
18. Kluczek, A.; Olszewski, P. Energy audits in industrial processes. *J. Clean. Prod.* **2017**, *142*, 3437–3453. [CrossRef]
19. Park, S.K.; Moon, H.J.; Min, K.C.; Hwang, C.; Kim, S. Application of a multiple linear regression and an artificial neural network model for the heating performance analysis and hourly prediction of a large-scale ground source heat pump system. *Energy Build.* **2018**, *165*, 206–215. [CrossRef]
20. Zhu, Y. Applying computer-based simulation to energy auditing: A case study. *Energy Build.* **2006**, *38*, 421–428. [CrossRef]
21. Wang, Z.; Wang, Y.; Zeng, R.; Srinivasan, R.S.; Ahrentzen, S. Random Forest based hourly building energy prediction. *Energy Build.* **2018**, *171*, 11–25. [CrossRef]
22. Sarker, I.H. Machine Learning: Algorithms, Real-World Applications and Research Directions. *SN Comput. Sci.* **2021**, *2*, 160. [CrossRef]
23. Koseleva, N.; Ropaite, G. Big Data in Building Energy Efficiency: Understanding of Big Data and Main Challenges. *Procedia Eng.* **2017**, *172*, 544–549. [CrossRef]
24. University of Melbourne. *Sustainability Report 2020*; The University of Melbourne: Melbourne, Austria, 2017; p. 42.
25. Kreider, J.F.; Claridge, D.E.; Curtiss, P.; Dodier, R.; Haberl, J.S.; Krarti, M. Building Energy Use Prediction and System Identification Using Recurrent Neural Networks. *J. Sol. Energy Eng.* **1995**, *117*, 161–166. [CrossRef]
26. Boussaada, Z.; Curea, O.; Remaci, A.; Camblong, H.; Meabet Bellaaj, N. A Nonlinear Autoregressive Exogenous (NARX) Neural Network Model for the Prediction of the Daily Direct Solar Radiation. *Energies* **2018**, *11*, 620. [CrossRef]
27. Azadeh, A.; Ghaderi, S.F.; Sohrabkhani, S. Forecasting electrical consumption by integration of Neural Network, time series and ANOVA. *Appl. Math. Comput.* **2007**, *186*, 1753–1761. [CrossRef]
28. Tian, W. A review of sensitivity analysis methods in building energy analysis. *Renew. Sustain. Energy Rev.* **2013**, *20*, 411–419. [CrossRef]

29. Bhattacharyya, B.; Jacquelin, E.; Brizard, D. A Kriging-NARX Model for Uncertainty Quantification of Nonlinear Stochastic Dynamical Systems in Time Domain. *J. Eng. Mech.* **2020**, *146*, 04020070. [CrossRef]
30. Hayter, A.J. Simultaneous Confidence Intervals for Several Quantiles of an Unknown Distribution. *Am. Stat.* **2014**, *68*, 56–62. [CrossRef]
31. Mikayilov, F.; Johnson, C.; Van Genuchten, M. Estimating Uncertain Flow and Transport Parameters Using A Sequential Uncertainty Fitting Procedure. *Vadose Zone J.* **2004**, *3*, 1340–1352.
32. Talebizadeh, M.; Moridnejad, A. Uncertainty analysis for the forecast of lake level fluctuations using ensembles of ANN and ANFIS models. *Expert Syst. Appl.* **2011**, *38*, 4126–4135. [CrossRef]
33. Li, S.; Foliente, G.; Seo, S.; Rismanchi, B.; Aye, L. Multi-scale life cycle energy analysis of residential buildings in Victoria, Australia–A typology perspective. *Build. Environ.* **2021**, *195*, 107723. [CrossRef]

Article

Towards a Blockchain-Based Peer-to-Peer Energy Marketplace

Yeray Mezquita [1,*], Ana Belén Gil-González [1], Angel Martín del Rey [2], Javier Prieto [1] and Juan Manuel Corchado [1]

[1] BISITE Research Group, University of Salamanca, 37007 Salamanca, Spain; abg@usal.es (A.B.G.-G.); javierp@usal.es (J.P.); corchado@usal.es (J.M.C.)
[2] Department of Applied Mathematics, Institute of Fundamental Physics and Mathematics, University of Salamanca, 37008 Salamanca, Spain; delrey@usal.es
* Correspondence: yeraymm@usal.es

Abstract: Blockchain technology is used as a distributed ledger to store and secure data and perform transactions between entities in smart grids. This paper proposes a platform based on blockchain technology and the multi-agent system paradigm to allow for the creation of an automated peer-to-peer electricity market in micro-grids. The use of a permissioned blockchain network has multiple benefits as it reduces transaction costs and enables micro-transactions. Moreover, an improvement in security is obtained, eliminating the single point of failure in the control and management of the platform along with creating the possibility to trace back the actions of the participants and a mechanism of identification. Furthermore, it provides the opportunity to create a decentralized and democratic energy market while complying with the current legislation and regulations on user privacy and data protection by incorporating Zero-Knowledge Proof protocols and ring signatures.

Keywords: blockchain; energy market; multi-agent system; negotiation; distributed ledger technology

Citation: Mezquita, Y.; Gil-González, A.B.; Martín del Rey, A.; Prieto, J.; Corchado, J.M. Towards a Blockchain-Based Peer-to-Peer Energy Marketplace. *Energies* **2022**, *15*, 3046. https://doi.org/10.3390/en15093046

Academic Editor: Mohamed Benbouzid

Received: 3 March 2022
Accepted: 19 April 2022
Published: 21 April 2022

Publisher's Note: MDPI stays neutral with regard to jurisdictional claims in published maps and institutional affiliations.

1. Introduction

The current traditional power grid is designed to transport energy over long distances. This characteristic of the traditional system implies that certain limitations exist, such as the maximum voltage capacity supported by the distribution lines [1]. When this capacity is exceeded, the heat generated by a line can cause it to sag or break, resulting in power supply instabilities such as phase and voltage fluctuations. Because the capacity of a line depends on its length and the transmission voltage, one solution is to create shorter lines and distribute the functionalities of the current power grid in smaller smart networks. These networks are called smart micro-grids, which are a type of discrete energy system that includes appropriated energy sources as well as power loads that provide power to residential, commercial, industrial, and governmental consumers. The main purpose of smart micro-grids is to provide affordable energy to areas independently of the main power supply network while optimizing the transmission of the energy.

In the current context of energy generation, thanks to renewable sources such as solar or wind, and together with the emergence of a new type of actor that consumes and produces energy within the system—the so-called prosumers—micro-grids have the potential to replace the traditional energy transmission system in the near future [2]. However, the rise of smart micro-grids comes with some challenges that must be faced. These challenges range from the vulnerability of platforms to DDOS attacks, to the emergence of intermediaries that do not contribute to energy distribution but end up making it more expensive [3].

In the past, some authors have proposed strategies for energy management on micro-grid platforms. For example, in [4], the excess or shortage of energy could be compensated by exchanging it with the utility grid or other external sources. However, that paper did not allow for direct energy exchange between individuals, nor did it allow for the automation and distribution of the platform. Without the use of blockchain technology, democratized energy markets could not be created. A blockchain network acts as a reliable distributed ledger that is governed by the platform and where information of value is stored. The network can be utilized to distribute the control and governance of the smart grid, along with the communication that is carried out within it, thus avoiding the single point of failure and eliminating those intermediaries that do not give any value to the platform. Moreover, blockchain technology (BT) provides a mechanism for protecting the actors against identity theft by signing direct communications between peers [5].

After studying the literature on this topic, it was found that none of the works had been able to propose a truly decentralized platform that enables peer-to-peer energy trading, automatically, and with dynamic prices. For this reason, we propose a distributed Multi-Agent System (MAS) based on blockchain technology to enable decentralized control over a micro-grid platform that allows for an automated exchange of energy between its actors. In the proposed system, the MAS manages the workflow of the micro-grid, e.g., the negotiations between peers in the local market, or the correct balancing of the energy network. By using a blockchain network, the control of the platform is distributed between the agents while the resilience of the communication channel between them is improved. This allows for the deployment of a platform without a single point of failure. Moreover, existing research works do not take into consideration users' anonymity and privacy, something that we would also like to tackle with the proposed framework.

This paper shows a thorough study of the most important features of blockchain technology and smart micro-grids in Section 2. Section 3 studies how previous works in the literature tackle the challenges of using blockchain technology in smart micro-grids. Furthermore, the section studies how automated negotiation between machines could be achieved and its viability. Section 4 describes the proposed platform, which is a combination of a MAS and a blockchain network for improved decentralization, as well as the security of the platform, along with the viability of the creation of a local automated energy market that optimizes the payoffs for the micro-grid stakeholders. Finally, Section 5 draws up conclusions and some final remarks on the conducted research.

Contributions

This work is relevant for designers, developers, and practitioners alike who are working in the field of energy distribution and renewable energy adoption and who will get the most benefits from the proposed framework. The main contributions of this paper are as follows:

- The design of a framework that will help developers to create new platforms that allow for the appearance of automatic peer-to-peer energy markets with dynamic prices.
- The proposed framework also provides user flexibility in the negotiation algorithms used. They will be able to implement the algorithm they want depending on their needs, with the only prerequisite being that the communications between agents follow the same ontology.
- The framework designed also provides anonymity to their users, complying with the current data regulations.
- Following the proposed framework, the future platforms developed and deployed will be more democratic and decentralized, thus eliminating the single point of failure.

2. Conceptual Foundations of Micro-Grid Platforms and Blockchain Technology

Traditional power grids deliver energy from a few central generators to a large number of consumers. This creates a closed market in which energy prices are dictated in a monopolistic way. Sometimes, to avoid abusive pricing by companies towards consumers, states need to implement regulatory measures, with the European Union [6,7] being an example in this case.

In the face of this monopolistic behavior, the literature has proposed the distribution of the traditional main grid into smaller micro-grids [2]. These micro-grids are comprised of a set of loads and generators. The set of generators can be composed of individual houses with solar panels on the roof. The entry of more entities into the energy market reduces the risk of oligopolies and avoids the intervention of states by imposing the use of regularization measures. This way, the energy market is converted into a more democratic market in which the offer and demand of energy will be the only factors that can regulate the energy price.

Micro-grid platforms make use of a great number of Internet of Things (IoT) devices that exchange crucial information between them. The continuous communication between the devices allows for the distribution of the management and control of any IoT platform. This helps with the optimization of the workflow of the system, but not without some drawbacks [8].

- Heavy reliance on exchanged messages. Since each part of the system is controlled by an independent entity, the other entities have to trust the messages received to understand the system's global state. If a malicious entity could somehow modify the content of those messages, the proper functioning of the entire platform would be compromised.
- Reliance on the truthfulness of the transmitted data. Entities of the platform have to rely on the fact that the data transmitted have not been tampered with by the sender entity to make an unfair profit. In addition, it is a possibility that databases will be attacked in order to steal, modify, or delete sensitive information about the entities that are taking part in the system's workflow.

In the literature, the use of BT has been proposed to overcome the listed flaws of this kind of platform. BT consists of a peer-to-peer (P2P) network of nodes, governed by a consensus algorithm that dictates how the information is stored within the network. This technology allows for the creation of a distributed ledger where anything of value can be stored.

The use of a blockchain network within any IoT system makes it possible to distribute the process workflow while eliminating other centralized entities [8]. In addition, by eliminating the single point of failure factor of centralized platforms, protection against some traditional forms of cyberattacks is gained. In this way, the blockchain is used as a bulletin where important information about the system is stored. Furthermore, the data stored within the blockchain network are kept in the same state after their storage, which means that the information is tamper-proof [9].

Within a blockchain-based system, a cryptographic mechanism of pairs of asymmetric keys is used, which signs and encrypts the data transmitted. Hence, as long as the blockchain network is big enough, the consensus algorithm keeps the information in a consistent state [10], the keys are not compromised, and the information transmitted and stored is secure from any attack, thus maintaining its integrity and authorship [11]. If this mechanism is also used in the exchange of messages between individuals of the system, then the messages are protected from being read and modified by unauthorized third parties [12].

A user needs to generate a random private key to make use of a blockchain protocol. This key is usually part of a cryptography mechanism that uses a key pair mechanism: the random private key mentioned and a public one derived from that. This public-key cryptography mechanism is used, not only because they allow for an efficient management of the keys, but also because it is impossible for an attacker to obtain the private key even when knowing the public one.

To interact with a blockchain protocol, an individual needs to generate at least one wallet address as an identifier. It is a three-step process, which starts with the generation of a random private key that only the owner should know. Then, through a one-way algorithmic transformation, the public key is obtained, which is shared with the network and is used to verify the signatures made by the user with their private key. Finally, the public key is hashed in order to obtain the wallet address to be used in the exchange of virtual assets between individuals within the blockchain protocol.

The process of exchanging assets is quite straightforward and shares the same steps as in every blockchain. Figure 1 illustrates how a user, Alice, wants to initiate a transaction with Bob with 2 coins. To do that, Alice signs the transaction (Tx) with her private key and broadcasts it in the network. Then, each node of the network verifies Alice's signature with her public key, and if the check is correct and the transaction is proven to have come from Alice, the network validates that she has the coins she wants to spend. If everything goes well, the transaction will be added to the blockchain.

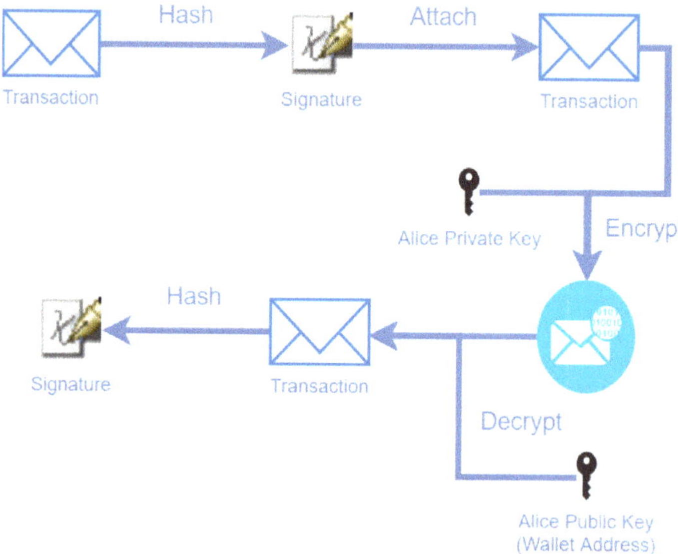

Figure 1. Example of the signature mechanism in a transaction.

2.1. Blockchain Consensus Algorithms

A consensus algorithm describes the mechanism that allows all agents in the system to coordinate in a distributed environment. It constitutes the only source of truth. Thanks to the consensus algorithm employed by the network of nodes, it is possible to keep the information stored and replicated in a consistent state. Among the functions of any consensus algorithm is ensuring that there is only one blockchain in the system, which can be an issue when a part of the network accepts a blockchain while the remaining nodes accept a different one (Fork). The consensus algorithm should enable the convergence of the chains into one as soon as possible. Moreover, it should offer resilience against attempts by malicious actors to take over the network and guarantee that there will not be any consensus failure when nodes try to add new blocks of data to the blockchain. Keeping the data stored in a blockchain makes it more difficult for attackers to take down the services of a system, and the attacker is forced to take down the majority of them to successfully hack the data [8].

There is a great variety of consensus algorithms, including the Proof of Work (PoW), Proof of Stake (PoS), and Practical Byzantine Fault Tolerance (PBFT) (see Table 1), or any of their variants that are the most widespread and have proven their effectiveness in practice [13].

Table 1. Comparison between consensus algorithms and their common usage.

Algorithm	Scalability	Consistency	Decentralization	Usage
PoW	No	Yes	Yes	Public blockchains
PoS	Yes	No	Yes	Public and permissioned blockchains
PBFT	Yes	Yes	No	Permissioned blockchains

PoW requires work to be performed by the miner and then verified by the network. The work required usually consists of the performance of a series of operations, algorithms, and mathematical calculations to be solved by the miners. These calculations vary and are different depending on the blockchain network they want to participate in. Each mathematical problem posed can only be solved by a very high computational calculation, which then encourages the nodes to behave in a certain way on the platform as compared with the simplicity of verifying the block mined. The greater problem of this algorithm is that a network using a consensus algorithm based on PoW wastes a massive amount of energy and is very slow. Therefore, it is not environmentally friendly and also not suitable for platforms that need to store information quickly [14].

PoS algorithms are based on the assumption that those who own more units of a PoS-based coin are especially interested in the survival and good functioning of the network that gives value to those coins. Therefore, they are the most suited to bearing the responsibility of protecting the system from possible attacks. That is why the protocol rewards them with lower difficulty in finding blocks (it is inversely proportional to the number of coins they prove to possess). The PoS algorithm has a theoretical vulnerability called the Nothing at Stake Theory, which has not been proven in practice. That theory states that forks in the blockchain network will occur more frequently [15].

In a PBFT consensus algorithm, all nodes communicate with each other, with the objective that honest nodes reach an agreement on the state of the system following the majority rule. Nodes not only have to verify that the message comes from a specific node, but they also have to verify that the message has not been tampered with. For the model to work, it is assumed that the number of simultaneous malicious nodes can never be equal to or greater than one-third of the total number of nodes. Therefore, the more nodes there are in the system, the more difficult it will be to reach that third. It is called practical in the sense that this proposal can work in asynchronous environments. This algorithm is used only in permissioned platforms and cannot be used in a public one, where nodes can access it freely [16].

In [17], the authors discussed the "blockchain trilemma", a term coined by Ethereum's founder Vitalik Butering to explain the problem of developing blockchain technology. According to this study, no blockchain satisfies the following three characteristics: scalability, consistency, and decentralization (see Table 1). For example, PoW solves the consistency and decentralization problems, but it lacks scalability. On the other hand, PoS can offer scalability and decentralization, but at the cost of consistency. Finally, PBFT-based algorithms can solve consistency problems while being scalable, but they centralize the process.

2.2. Blockchain Accessibility

The implementation of blockchain technology in the real world depends on the accessibility of the network underlying this kind of platform. If a player needs permission to be part of the blockchain network, it is said that it is a permissioned one. These kinds of networks are used in platforms where the actors are known, although they each have different interests. On the other hand, if anyone can be part of the network without requirements, the network is called a public blockchain.

A public blockchain, based on PoW, is less efficient in terms of reaching consensus and therefore managing transactions per second because it offers a truly decentralized ecosystem with proven security against attacks, with Bitcoin and Ethereum being their main representatives [10] (see Table 1). Public blockchains that make use of another consensus algorithm, such as PoS or any of its variants, are far more efficient, although they lose some consistency. Blockchain networks that use PBFT-based consensus algorithms could only be used in permissioned environments because they lose decentralization in favor of scalability; to have consistency, it is required that the actors are known.

2.3. Smart Contracts

Another relevant aspect of some blockchain technologies is that they allow for the deployment and execution of coded scripts called smart contracts. Those scripts, due to the immutability feature of the blockchain technology, are considered self-enforcing and are used to automatize some processes, such as payments between entities within a platform that would otherwise need human intervention and/or that of third parties [18]. The code of smart contracts is transparent to the players that can make use of it, which means that they know the programmed clauses that rule it. Then, when those parties agree to use a smart contract, the workflow of the interactions between them is governed by the rules coded in the smart contract, all without the need for human hands to verify the process [19]. A smart contract ensures that the agreement will be carried out automatically when the conditions agreed upon are met [20].

3. Related Work on Micro-Grid Platforms Based on BT

Blockchain technology has been used to improve the performance of a broad range of platforms in today's industries. The state of the art encompasses, to enumerate a few examples, the pharmaceutical industry [21,22], the agri-food sector [23,24] as well as healthcare [8,25–27] and education services [28–30].

In this section, we will detail a small study on the state of the art related to the use of blockchain technology in the field of micro-grid platforms. Then, it will be followed by a study on the automatic negotiation algorithms that have been proposed in order to understand the requirements that need to be implemented in this type of platform.

3.1. Blockchain Technology and Micro-Grid Platforms

In the literature, we found some works that discussed the use of blockchain-based micro-grid platforms to create energy markets, focusing on specific characteristics such as the use of cryptocurrencies, decentralization, security, privacy, and state estimations [12,31–34].

Pichler et al. [35] studied real-world use cases of platforms based on blockchain technology and whose aim was to allow for the direct exchange of energy between its actors. The platforms studied have a general common aim: to create local markets based on renewable energy communities. However, they share the same cons: they do not try to create an autonomous market, and they do not offer real anonymity and privacy to their users (see Table 2).

A working example is the Pylon network [36], a Spanish startup that makes use of its permissioned blockchain-based Litecoin technology combined with a smart meter to certify energy flows and enable virtual transactions with the use of their own token. It makes use of a Proof of Cooperation (PoC) consensus algorithm, and its main aim is to create a neutral database, one that is not governed by the companies that sell the energy,in order to help the user decide how to optimize the energy costs. They made their platform open source to receive help from the community in case any kind of improvement is needed for their network. In Slovenia, SunContract [37] has created a market for peer-to-peer transactions of energy based on BT. They launched a crypto-asset within the Ethereum network in order to use it in the exchange of energy between the entities that are participating in the platform. On the other hand, there is Enosi [38], an Australian company whose aim is similar to that of SunContract: to create a community of peers transacting energy directly between them. By using smart metering, they trace, match, and settle energy production and consumption. Because of the platform, the producers can directly offer a price to the end consumer, with cheaper prices instead of the artificial ones that the power oligopolies have in the traditional energy market.

In the case of the Brooklyn micro-grid [39], LO3 developed the TransActive Grid elements (TAG-e), which allows for the exchange of energy between peers, the balancing of the grid, or the emergency management of the network. A TAG-e is composed of two elements: an electric meter and a computer. They are meant to read the information on the state of the grid and share it with other TAG-e in order to act upon the collected information. The market created with this platform allows for the trading of energy between peers with fixed prices; however, automatic negotiations within it are not permitted.

Table 2. Comparison of the studied startups.

Project	Description	Pros	Cons
Pylon Network [36]	The main aim is to create a neutral database. Makes use of its permissioned blockchain-based Litecoin technology. It makes use of a Proof of Cooperation (PoC) consensus algorithm. In addition, a smart meter (METRON) certifies energy flows and enables virtual transactions using their own token.	• Open source • Scalable • Latency • Improve prices	• Nothing about user's data privacy • It is not designed to create an autonomous market
SunContract [37]	The main aim is to create a marketplace that allows customers to trade energy without the need for intermediaries. They managed a market for P2P energy transactions based on BT for more than 2 years. They enable virtual energy transactions using their own token.	• Scalable • Latency • Improve prices	• Nothing about user's data privacy • It is not designed to create an autonomous market
Enosi [38]	Their main aim is to create a marketplace that allows the energy customers to trade energy without the need for intermediaries. They certify energy flows via smart metering.	• Scalable • Latency • Improve prices	• Nothing about user's data privacy • It is not designed to create an autonomous market
Brooklyn Micro-grid Network [39]	This project created a local energy marketplace in Brooklyn. Bacuase of it, prosumers can trade their energy surplus with their neighbors.	• Scalable • Latency • Improve prices	• Nothing about user's data privacy • It is not designed to create an autonomous market

3.2. Negotiation Algorithms on BT-Based Micro-Grid Platforms

Due to the increase in the production of renewable energy, grid consumers need to be flexible in adjusting their energy consumption. This adjustment occurs through different demand response mechanisms: either by reducing electricity consumption during hours where the global consumption is at its highest (peak hours) or by discouraging the consumption during those hours by affecting the prices with financial incentives. Regarding the last demand response mechanism, it is possible to implement it in a negotiation process based on the law of supply and demand. Then, peers of the network can trade energy directly through a local P2P network, thus allowing for the movement of local funds within the local economy [40] (see Figure 2).

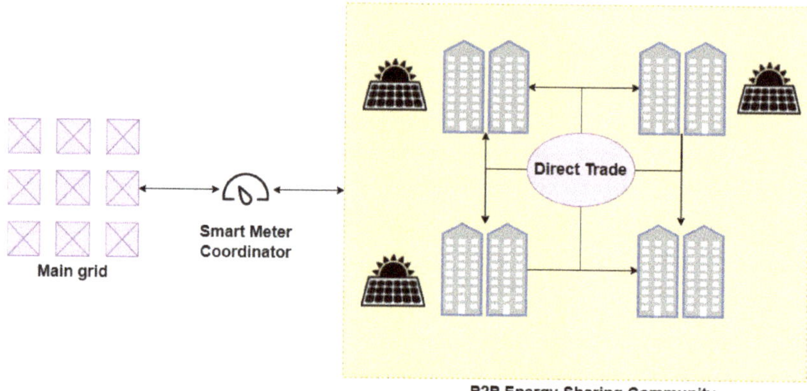

Figure 2. Basic diagram of a micro-grid architecture. The actors of the P2P network exchange energy locally and can potentially sell energy outside the community thanks to the existence of a smart meter coordinator.

In the study by Long et al. [41], in order to optimize the energy prizes for the players participating in the micro-grid community, they proposed a non-linear programming optimization algorithm with a rolling horizon of 30 min. The method proposed made use of a model based on the supply and demand proposed in [42]. In this model, the prices of energy fluctuate through the day, with a constraint that the price of the energy generated within the micro-grid should never be higher than the price of the energy bought from outside the grid. Moreover, the prize of the energy sold to the external grid must always be higher than that of the energy sold within the micro-grid. In this work, it was found that the smart meter coordinator was the most vulnerable part of the architecture proposed. The possibility of the smart meter being hacked was not considered, since BT was a mechanism that was required to avoid this kind of vulnerability while distributing the control of the activities and the negotiations carried out within the community.

Authors of [43] modeled a micro-grid scenario in which two variants appear, one based on cooperation between the different actors on the platform, and the other in which the actors play more selfishly in the market. This platform is only viable when all costs are equally shared between all households; therefore, there is no automated negotiation between the peers of the platform proposed. In this paper, it was also stated that a real scenario wherein all the actors collaborate is not feasible.

The companies studied in Section 3.1 allow for the trading of energy at a fixed price given by the producer. In other theoretical works such as that by Noor et al. [44], to allow for the exchange of energy with dynamic prizes, a game theory-based model was proposed. In this work, the blockchain network was used to distribute the control of the platform along with the exchange of information as a transparent energy market was created. Here, the actors that formed part of the platform negotiated the price of the energy in an automatic manner using a non-cooperative game-based algorithm to optimize their payoffs, based on the energy load of the entire grid. An important downside of this approach is that the system must know which specific appliances are connected throughout the entire network; this, along with the fact that no mechanism of encryption is used to store the data within the blockchain, creates a great privacy problem for the users.

3.3. Literature Review Conclusions and Manuscript Objectives

The present paper aims to create a transparent energy market with proven distributed security. Because of the nature of this kind of system, it will be impossible to use a public blockchain such as Ethereum due to its high fees and slow speed. A consortium of peers in the micro-grid will be needed to create a permissioned network that makes use of a protocol such as PBFT, PoS, or dPOS, where it is assumed that all the network's participants are known and have the common goal of wanting the platform to work.

One of the features that the companies studied are lacking is the use of an automatic negotiation between the players of each platform. Our model makes use of a non-cooperative game between the consumers and producers that will regulate the energy market price in an autonomous way, thus allowing the stakeholders optimize their payoffs. Our model and its interactions with the blockchain are thoroughly explained to help startups and entrepreneurs to develop this kind of system. Furthermore, the scenario proposed in our work is based on a rolling horizon such as the one proposed by Long et al. [41], but with a time window of an hour to make the transactions more viable in the actual Ethereum network.

In the literature, compliance with the General Data Protection Regulation (GDPR) [45] has been found to be an important issue. Because of this, a careful selection of which data are to be collected and stored in the public ledger is needed, as well as which data must be encrypted and hidden from unauthorized peers. Moreover, ensuring the integrity and authenticity of the data is required by protecting it and the communication channels from unauthorized users [35]. Another issue of the proposed platform is its heavy dependence on the legislative framework of the country where it is to be installed. Laws that regulate the transaction of renewable energy between peers within communities are needed, such as in the case of Belgium, Greece, and Germany [35].

4. Proposed Architecture Design

In this section, we will describe the design of the proposed architecture, which aims to: (i) decentralize the energy market, (ii) automate, as much as possible, the energy market in small communities, (iii) and provide anonymity and privacy to its users. In the literature studied in the Section 3, there are working proposals that meet some of the above-mentioned requirements, although not all of them together.

The proposed architecture will follow the paradigm of distributed Multi-Agent Systems (MAS), which, in combination with blockchain technology, allows for the distribution of the processes and the control of the platform. The use of multi-agent systems was chosen because other works successfully achieved their main objective with the use of this paradigm, with the optimization and decentralization of platforms of any kind [46]. In the proposed architecture, features from different works studied in the literature have been put together, thus enabling decentralized control over the platform without a single point of failure and allowing for a negotiation process between the peers of the network as well as complying with the GDPR.

- Through the MAS, the control and management of the micro-grid platform is achieved, along with the negotiation between peers for energy in the market. However, to achieve full decentralization of the platform, the use of a blockchain platform is required, in which the smart contracts deployed will be used by the agents in the workflow of the platform. Thanks to this approach: (i) we will achieve a decentralized platform without a single point of failure; (ii) we will provide confidence to platform users and agents that agreements would be enforced, and in case they are not, encourage trust that the platform will compensate those who behave while punishing those agents who do not; and (iii) we will allow for the optimization of the prices of the energy transacted within the platform, balancing them while maximizing the payoffs of each kind of actor involved.
- The smart meters read the energy consumed and/or produced by each household. They are connected to each independent house, representing a peer in the micro-grid network. Each smart meter is connected to the internet and interacts with the blockchain on behalf of the household. Moreover, the agent who negotiates with their peers to buy or sell energy should be deployed here or in a device connected to the smart meter.
- The use of a blockchain network allows for the distribution of the communication and the interactions between the agents of the platform. The network is used not only as a

historical log in that each agent stores their activity on the platform, but it is also used as a validation and tamper-proof system that will help them to trust the platform and the activities of the actors involved. In addition, the smart contracts deployed in the network help in the control of the workflow of the platform.

- The information stored in the blockchain is encrypted, maintaining the data hidden from others. It is possible to maintain a verified and encrypted log in the blockchain by using Zero-Knowledge Proof (ZKP) protocols. Furthermore, by using ring signatures, the identity of the entities that store information within the blockchain is kept secret.

4.1. Blockchain Technology and Smart Contracts

The design of the proposed platform is based on the negotiation, payment, and exchange of energy. In time windows of one hour, agents negotiate the energy prices based on the amount they wish to transact during the following hour. The platform will use a permissioned blockchain, governed in a consortium way between the market actors. A permissioned blockchain network, as seen in the background section, allows for a high transaction output but with a very low cost.

The use of a permissioned network is proposed because it achieves two things that cannot be achieved using a permissionless network [47]: (i) transaction speed, since there are only known nodes within the network, it is possible to make use of faster consensus algorithms at the cost of a certain level of security; (ii) system scalability, because of the above-mentioned characteristics, by not requiring a large computational capacity to reach consensus, the system is scalable; (iii) the network protocols can be adapted to the system requirements during development, e.g., with the addition of ZKP protocols and ring signatures that are not available in any permissionless blockchain that allows for the deployment of smart contracts.

In a consortium blockchain, only verifiable actors are allowed to take part in the proposed platform. If new actors, e.g., new households, want to take part in the created market within the micro-grid, they have to make a proposition to the platform; here, in a consortium and not in an automated way, the actors of the micro-grid will vote if they will let them enter or not. If the actors suspect that the new actor trying to enter the platform has no good intentions or has intentions that are not aligned with the well-being of the platform, they will not be allowed to enter. On the other hand, if it is a typical household that wants to benefit from the good use of the platform, they will allow it to enter. To summarize, the consortium blockchain proposed in this framework should be governed equally by all the nodes of the blockchain; they all have equal voting rights.

Due to the characteristics of the blockchain network used, the platform will need a margin to store the agreements carried out by the agents. In the proposed platform, the margin is 5 min, enough time for the network to validate the information of the platform [48]. In that time window of 5 min, agents cannot continue their negotiations. Then, after 55 min, the agents will only sign the agreements that best benefit them after the negotiation period. In this way, the energy prices are fixed by each batch of energy independently negotiated, dependent only on the supply and demand, and each buyer and seller will make their own decisions based on their situation and the payoffs they want.

On the platform, smart contracts are used to generate tokens that represent the amount (in KWh) of energy available for exchange in the batteries. The virtualization of this energy is achieved by using Ethereum's fungible token standard: ERC20. Making use of standards is important for future system extensions as well as for the improvement of interoperability.

In the proposed platform, a smart contract is used to control the workflow of the platform (see Figure 3). The usual sequence of steps followed by the platform is described below:

1. Through the function *PublishInfo()*, agents can identify themselves on the platform. They can store data in relation to how other agents can initiate negotiations with them, the household they belong to, etc. With that information, it is possible for authorized actors to carry out auditory processes as well as to track their activity on the platform. This step should be performed the first time an agent is deployed in the system.

2. To publish any energy offer on the platform, authors should call the function *MakeOffer()*. Agents can calculate the forecasted energy surplus that could be sold to the network and create an offer with the predicted amount of energy for the next time window.

3. When an agent predicts a need to buy energy for the next time window, it will need to call the function GetOffers(). This function will return all the information related to the offers published for the next time window. Then, the agent will start the negotiation process directly with all the publishers of offers.

4. During the negotiation process, the agents try to reach an equilibrium on the price of the energy and the amount that must be bought. The price of the energy sold has an upper constraint, which is the price of the energy bought from outside the grid. It also has a lower constraint, which is the minimum price needed to produce the energy. Between those thresholds, agents have the autonomy to decide; they could use whatever negotiation algorithm they find more comfortable with as long as it exchanges messages following the ontology defined by the platform communications. The agents negotiate on the basis of different parameters such as the energy needed to buy or sell, the time left to finish the negotiation, the number of buyers or sellers, the amount of energy to be expected to generate or consume in the next time window, etc. When the last minutes of the negotiation are reached, each seller agent will start agreeing to sell the energy to those that offer the higher prices until the energy is all sold out. The buyer agents will do the opposite—they will buy at the lower prices given by the sellers during their negotiation. Because of the constraints, it is ensured that all the energy will be sold out; no buyer will buy from the main grid while energy is still available within the system. Therefore, each agent should have a time out to get answers from an offer. If they do not receive an answer during that time out, they will have to drop the offer and try to reach an agreement with the next best offer on their list. This will ensure an equilibrium point while avoiding getting stuck in infinite waiting periods.

5. After negotiating the price and the amount of energy to sell and to whom, the seller can publish on the blockchain to whom, how much, and for how much they are selling the energy with the function *AllowTransaction()*.

6. Finally, when the corresponding smart meters have detected the flow of energy to and from a house, automatic payments can be made by calling the function *MakeTransaction()*.

For the platform to function optimally, the amount of energy available to exchange within the market must be auditable. A guarantee is needed that this energy exists on the platform, so the smart meters in charge of reading this energy from the batteries and virtualizing it into tokens for sale undergo periodic auditing processes to ensure its proper functioning [47]. In addition, the smart meters are in charge of reading the energy flow in and out of the houses, another critical element for the proper functioning of the platform. In this sense, we call smart meters oracles, since they are in charge of virtualizing real-world data in smart contracts, a critical point of any platform based on blockchain and of which it is necessary to be very cautious [49].

Each agent of the platform interacting in one way or another with the blockchain network needs to make use of a wallet. Some agents, such as those in charge of using virtual money in exchanges, need to obtain that money beforehand. For example, if a household consumes more energy than it produces, it will need to put fiat money on the platform; hence, a consortium of human agents and/or machines will be needed to virtualize the money introduced and mint more tokens that represent it. In addition, they must allow the withdrawal of real money when a user decides to take out part of the virtual money for use outside the platform.

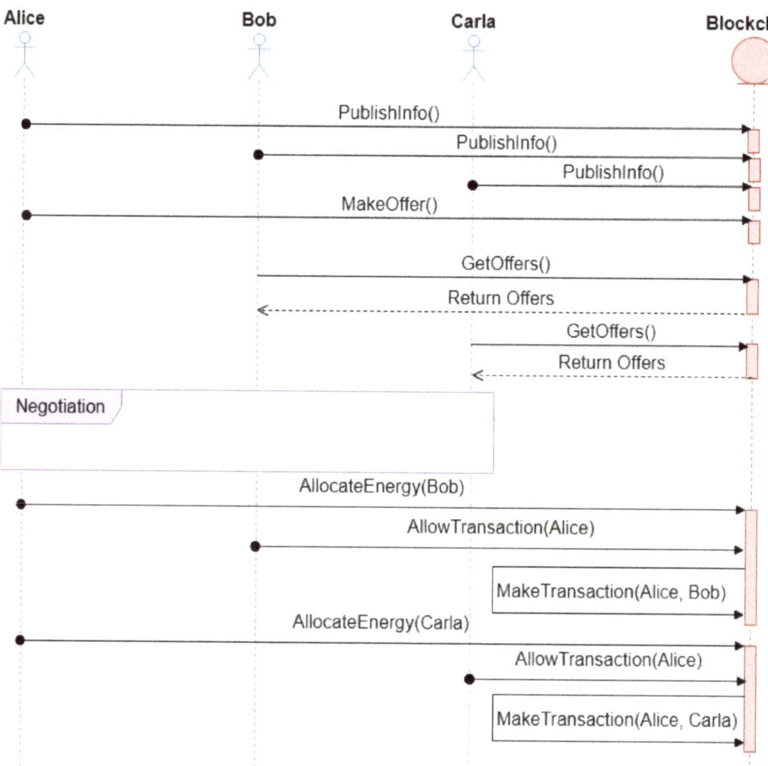

Figure 3. UML diagram of a sample workflow of the proposed platform.

4.2. Privacy Preservation Protocols

One of the problems that arise in state-of-the-art literature is that of user privacy and data protection, which are needed to comply with the GDPR. In this regard, the proposed model has been designed using protocols based on ZKPs used by the Monero cryptocurrency and described in [50]. Thanks to the use of these protocols, it is possible to hide the users who perform transactions within a blockchain as well as the related information [51].

For example, in [52], the authors proposed a framework that allows people who have been in close contact with infectious disease patients to be traced. Moreover, the authors proposed the use of ZKP to protect patient information based on bulletproofs [53].

The ring signatures protocol is used to allow actors to call smart contracts anonymously. This protocol requires what is called a Key Image [51], obtained from a list of randomly selected public keys (see Figure 4). The public key of the actor performing the transaction is also required since the transaction must be signed. Given that all the selected keys have the same probability of performing the transaction, it is not possible to associate the transaction with the real user. In addition, these groups of actors are improvised randomly from the pool of transactions.

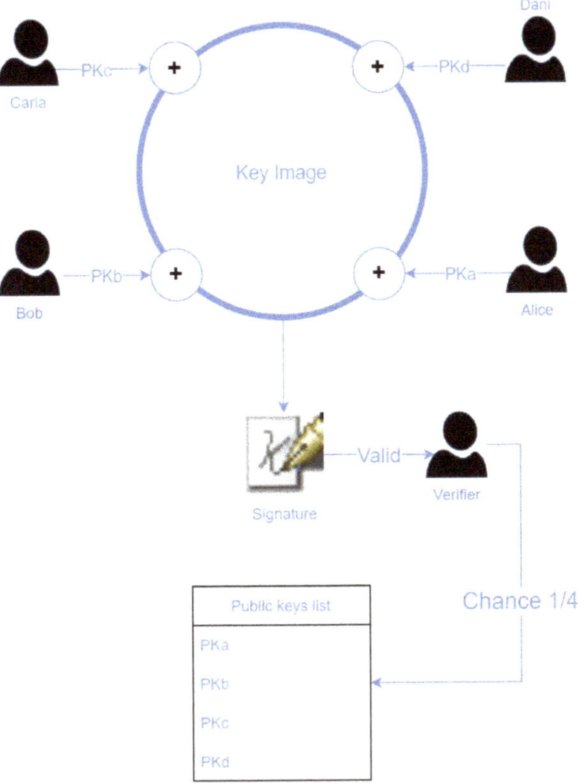

Figure 4. Key image, created from a list of the signatures of the users Bob, Alice, Carla, and Dani.

To complement the ring transaction signing process and ensure the anonymity of the actors within the system, stealth addresses are used in the smart contracts to identify actors. It is impossible to link these addresses to a user; however, a user can identify the stealth addresses that belongs to it. Taking advantage of the properties of elliptic curves [51], a stealth address (P) is defined by Equation (1):

$$P = F + S \tag{1}$$

where F is defined in (2), and S is the public key of the recipient of the transaction.

$$F = Hash(rS) * G \tag{2}$$

where r is a random private key generated by the actor that emits the transaction, and G is the base point of the elliptic curve.

To identify which stealth address belongs to a user, thanks to the properties of Equation (3), the actor can hash the product of the address public key (R) and its private key (s), then the public key (S) has to be added, and the final result is the stealth address. For a user to prove that a stealth address belongs to them, they need to recover the one-time private key generated for that transaction. By hashing the product of the stealth-address public key (R) and the user private key (s), and then adding the private key s to the obtained hash, it is possible to recover the one-time private key of the address (r). Then, it is necessary to sign the transactions from that address with that key to prove the ownership.

$$rS = rsG = rGs = Rs \tag{3}$$

where R is the public key of the randomly generated private key, and s is the private key of the transaction recipient.

The use of these protocols increases the need for the computational power of each actor that makes use of them. In Figure 3, it is possible to study the number of times each agent must write to the blockchain, thus making use of these protocols, within a system based on the proposed framework. The agents only need to write the information in the following cases:

1. When they are registered within the system and store information related to them. In the whole life cycle of the platform, this occurs once for each agent.
2. At the end of each hour, every agent writes in the blockchain the agreement reached during the negotiation process. For example, if Alice reaches an agreement with Bob and Carla, then Alice will need to create two transactions. On the other hand, according to the example, Bob and Carla only need to create one each.

This step will depend on the number of agents involved, but with the time limits proposed—from 5 to 10 min—in a permissioned blockchain, it is enough time to not overcharge the agents and their computational resources. Therefore, the performance will not be affected when the system escalates.

4.3. Security Model

This section details the security assumptions made by the framework and how the data generated within the platform are treated. The implementation of blockchain technology in this platform ensures the application of a secure identification protocol between the actors. Furthermore, the information stored is tamper-proof, and the smart contracts deployed allow for the decentralized control of the platform, ensuring that there will not be a single point of failure that will be prone to attacks.

Regarding the storage of the generated data, which will be used to create the predictive models that will help with the proper functioning of the system, each actor will be responsible for them. We assume that each actor is responsible for providing an access point to their data so that they can control to whom they give access to the data. For each hour, a batch can be created with the generated data, storing in the blockchain a hash of such data that will help to verify that it has not been modified afterwards. The use of auditability systems allows for the generation of data that can be trusted. Otherwise, it would be impossible to know that the generated data has not been modified before the storage of its hash in the blockchain [47].

As for the proposed privacy protocols, they ensure that the information stored in the blockchain cannot be read by third parties without permission, nor will it be possible to identify or track user activity. This information that is stored is, for example, the energy bids posted, the money paid for energy transactions, or the amount of energy transacted. The only vulnerability of this platform is when an attacker steals the keys of a user. However, this is not possible just by using the platform; it can only happen if the user is not careful enough with the passwords used or with where the keys are stored.

In this work, we have made security assumptions that the network of blockchain nodes is large enough so that it is not easy to throw it from a typical Distributed Denial of Services (DDOS) attack. It has also been assumed that the actors that are part of the platform benefit more from its proper functioning than from trying to sabotage it. Different actors could collude with each other to achieve a greater benefit, but this scenario is not realistic based on the study conducted by [43]. Having in mind the previous assumptions, we can thus say that the actors of the platform will benefit from the creation of this platform and the competition between them rather than in trying to sabotage the negotiation process and the well-being of the platform.

4.4. Multi-Agent System

This section will describe the MAS structure in detail. It is divided into four different subsystems, in which the agents are grouped according to their function within the platform (see Figure 5). The following is a detailed description of the different subsystems and the agents that comprise them:

- Device Driver System (DDS). This system groups all the agents in charge of the management and control of the different smart devices of the platform (e.g., batteries, smart meters, PV panels). These agents are allowed to interact with the blockchain network, so they also have an assigned wallet to identify and track their activity within the platform, thus helping in the auditing process. The agents in charge of monitoring the state of the PV panels (e.g., their energy production, the provided voltage and current, and their active and reactive powers) are the PV agents (PVA). They store those data in the blockchain, which helps their owners to monitor them while also owning that information which they could sell in the future. The batteries are monitored by agents called Battery State Agents (BSA). They store in the blockchain data related to the state of a battery, its charge and discharge capability, and its current state of charge. The agent that stores the data related to the flow of current from or to a household is the Smart Meter Agent (SMA).

- Micro-grid Operator System (MGOS). In this system, all those agents that are responsible for monitoring, controlling, and managing the status and good credit of the micro-grid are grouped together. These agents are also connected to the blockchain, storing the relevant information that favors the traceability of the micro-grid monitoring, flows of power to and from the utility network, the balance of the micro-grid power, and the voltage level (Micro-grid Operator agent or MGO), or the energy transactions made from the grid to the micro-grid and vice versa (External Market Interactor Agent or EMI). In addition, this system owns a series of batteries that improve the balance of the grid load, governed by the State Of Charge agents (SOC). This part of the platform is economically maintained by the penalties of users who do not fulfill their part of the contracts and by the exchange of energy between the external grid and the micro-grid.

- Data Analytic System (DAS). This system is crucial for the platform as it is in charge of grouping all those agents that are in charge of the data market and the creation of predictive models, which are required by the rest of the agents of the system to be able to infer the amount of energy they expect to obtain in the next hour, that which they could sell, and that which they will need to buy based on their past consumption. The agents in charge of reading the data provided by the other subsystems of the platform on the blockchain and merging it with data coming from other external data sources are called Data Reader Agents (DRA). The agents that create and update new behavioral models on demand are the Knowledge Extractor Agents (KEA). The agents that make predictions based on these models and the information extracted from the environment are the Forecasting Agents (FA). This subsystem benefits from the data market created with the addition of blockchain technology to the platform. As it has been found in other works in the literature, it is also possible to improve the creation of the models with the use of blockchain technology by applying a federated learning framework similar to the one proposed in [54].

- Transaction Manager System (TMS). In this subsystem, all those agents that are responsible for the negotiation and exchange of energy within the micro-grid are grouped. These agents make use of the blockchain network to publish and search for energy offers as well as register the agreements that take place. The agents in charge of publishing the offers are the Seller Agents (SA), while those who search for them in order to buy are the Buyer Agents (BA). The agents in this system negotiate with each other directly and make use of the DAS to estimate the energy they will need to buy and/or sell. As a way to improve the search process in the blockchain, a middleware layer could be used to optimize the search for information (offers in this case) within the blockchain, such as the one proposed in [55].

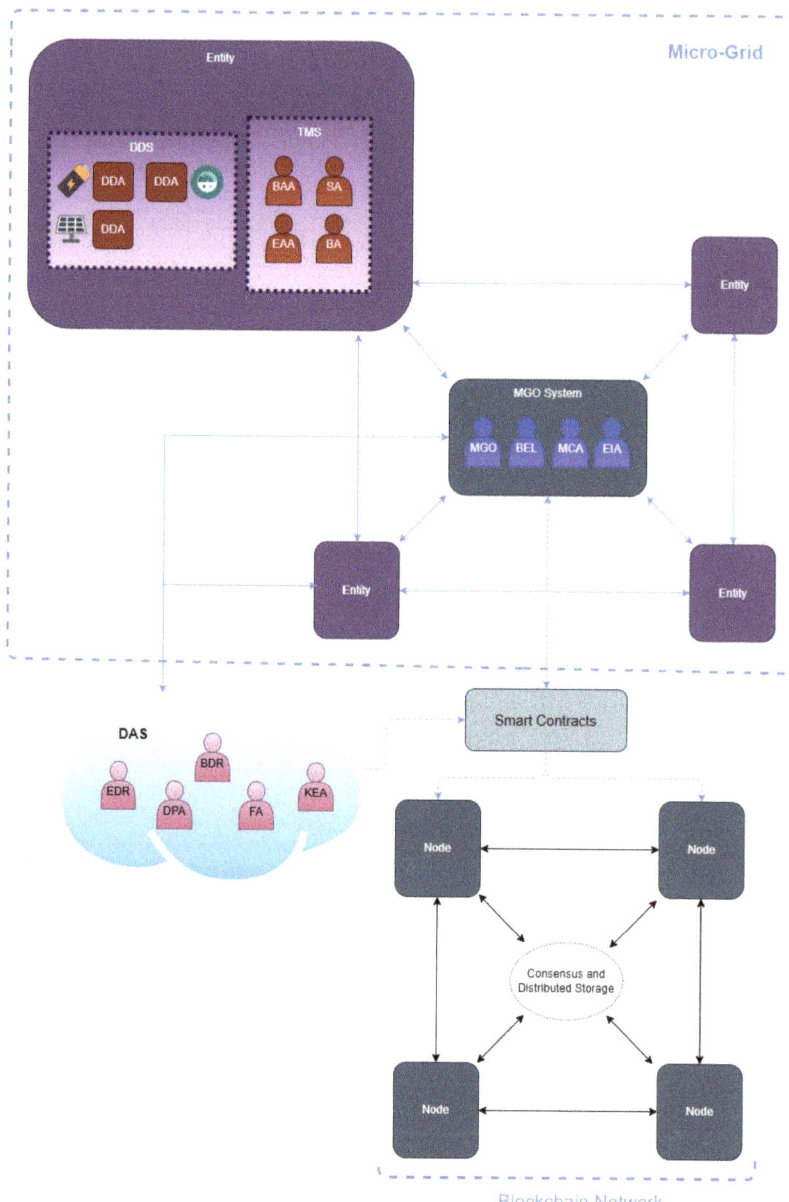

Figure 5. Proposed platform architecture.

4.5. Deployment of the Platform

In the proposed platform, there are three types of actors: consumers who receive energy that they buy from the grid, PV panels as producers that produce the energy and dump it into batteries, and batteries as prosumers who store and distribute the energy (e.g., consumed by their owners, or, if leftover, dumped into the micro-grid to make a profit). There are also actors who are in charge of the good credit of the micro-grid and who additionally make a profit by connecting it to the external grid. Finally, there are other players who help the platform to function properly, such as those in charge of exchanging fiat money for digital money and vice versa.

As shown in Figure 6, the interconnections between the parts of the platform are made through the Internet, in the creation of cloud services. The platform agents in charge of knowledge extraction with regard to the platform data needs large computational power; hence, the infrastructure is outsourced to a provider (e.g., Amazon Web Services, Google Cloud, or Azure). The blockchain network, controlled by the platform actors, is accessible to the parties and does not need high computational power; only one computer per participant is required to be always on. The agents in charge of controlling the smart devices will need to be deployed in them or in a system such as a Raspberry Pi that has direct access to them. The rest of the agents only need to be deployed in computers that always have access to the Internet and do not have any special requirements.

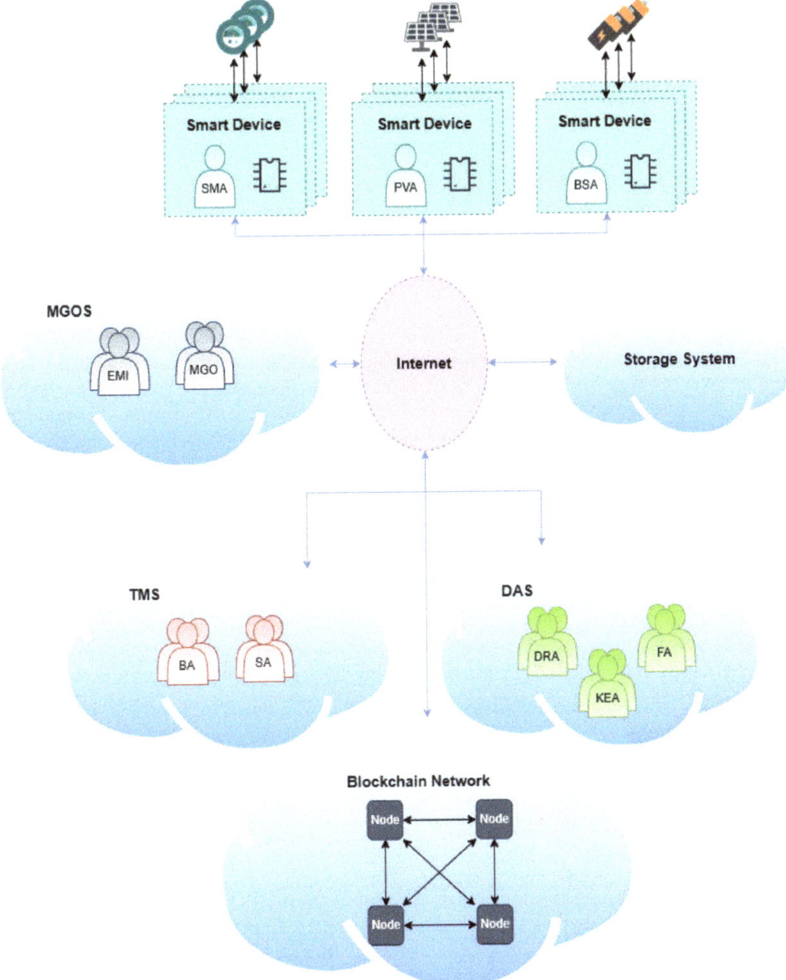

Figure 6. Platform deployment diagram.

5. Conclusions and Future Work

This manuscript highlights and discusses concepts and technologies such as blockchain, smart micro-grids, and negotiation algorithms that have been applied to the energy market. Furthermore, this work elaborates on the proposal of an innovative high-technology-based architecture of a fully distributed and autonomous smart micro-grid to support new forms of business in the smart energy market. With independent and dynamic pricing between the transactors in the network, the proposal presented makes it possible to create a Local Energy Market (LEM) in order to achieve efficiency in the transmission and distribution of energy as compared to the traditional distribution model.

In the mentioned context, this paper has studied the implementation and development of smart micro-grids that would allow for the entrance of more entities, whose aim is self-consumption and making money with the excess energy generated, as competitors in the energy market. If it is possible to introduce more actors into the power market that can compete for revenue in the energy market, the regularization of prices is no longer necessary. Because the energy transactions between entities of a micro-grid that are closer to each other are cheaper and better than those made between distant entities, the market law of supply and demand would work, thus making their current regularization unnecessary. Finally, the use of ring signatures and ZKP protocols has been proposed to ensure the privacy of the users and the protection of the data stored within the platform, thus complying with the GDPR.

In general terms, for future work, the designed proposal should be implemented as a pilot project. This will allow for the design of ad hoc consensus algorithms for the energy market. In this way, it will be possible to validate the proposed framework as a standard guideline for similar platforms. This will help to reduce development costs and encourage the adoption of the system by companies and individuals.

Author Contributions: Funding acquisition, J.P. and J.M.C.; Investigation, Y.M.; Methodology, Y.M. and A.B.G.-G.; Project administration, Y.M.; Supervision, A.B.G.-G., A.M.d.R., J.P. and J.M.C.; Writing—original draft, Y.M. and A.B.G.-G.; Writing—review & editing, Y.M., A.B.G.-G. and A.M.d.R. All authors have read and agreed to the published version of the manuscript.

Funding: The research of Yeray Mezquita is supported by the pre-doctoral fellowship from the University of Salamanca and co-funded by Banco Santander. Besides this work has been partially supported by the Institute for Business Competitiveness of Castilla y León, and the European Regional Development Fund under grant CCTT3/20/SA/0002 (AIR-SCity project).

Institutional Review Board Statement: Not applicable.

Informed Consent Statement: Not applicable.

Data Availability Statement: Not applicable.

Conflicts of Interest: The authors declare no conflict of interest.

References

1. Hirst, E.; Kirby, B. *Transmission Planning for a Restructuring US Electricity Industry*; Consulting in Electric-Industry Restructuring: Washington, DC, USA, 2001.
2. Fang, X.; Misra, S.; Xue, G.; Yang, D. Smart grid—The new and improved power grid: A survey. *IEEE Commun. Surv. Tutor.* **2011**, *14*, 944–980. [CrossRef]
3. Memon, A.A.; Kauhaniemi, K. A critical review of AC Microgrid protection issues and available solutions. *Electr. Power Syst. Res.* **2015**, *129*, 23–31. [CrossRef]
4. Bui, V.H.; Hussain, A.; Kim, H.M. A multiagent-based hierarchical energy management strategy for multi-microgrids considering adjustable power and demand response. *IEEE Trans. Smart Grid* **2016**, *9*, 1323–1333. [CrossRef]
5. Tosh, D.K.; Shetty, S.; Liang, X.; Kamhoua, C.; Njilla, L. Consensus protocols for blockchain-based data provenance: Challenges and opportunities. In Proceedings of the 2017 IEEE 8th Annual Ubiquitous Computing, Electronics and Mobile Communication Conference (UEMCON), New York, NY, USA, 19–21 October 2017; pp. 469–474.
6. Cameron, P.D.; Brothwood, M. *Competition in Energy Markets: Law and Regulation in the European Union*; Oxford University Press: Oxford, UK, 2002.
7. Von Danwitz, T. Regulation and Liberalization of the European Electricity Market-A German View. *Energy* **2006**, *27*, 423.

8. Mezquita, Y.; Casado-Vara, R.; González Briones, A.; Prieto, J.; Corchado, J.M. Blockchain-based architecture for the control of logistics activities: Pharmaceutical utilities case study. *Log. J. IGPL* **2021**, *29*, 974–985. [CrossRef]
9. Liang, X.; Shetty, S.; Tosh, D.; Kamhoua, C.; Kwiat, K.; Njilla, L. Provchain: A blockchain-based data provenance architecture in cloud environment with enhanced privacy and availability. In Proceedings of the 17th IEEE/ACM International Symposium on Cluster, Cloud and Grid Computing, Madrid, Spain, 14–17 May 2017; pp. 468–477.
10. Buterin, V. Ethereum: Platform Review. In *Opportunities and Challenges for Private and Consortium Blockchains*; Available online: http://www.smallake.kr/wp-content/uploads/2016/06/314477721-Ethereum-Platform-Review-Opportunities-and-Challenges-for-Private-and-Consortium-Blockchains.pdf (accessed on 19 April 2022).
11. Huh, S.; Cho, S.; Kim, S. Managing IoT devices using blockchain platform. In Proceedings of the 2017 19th International Conference on Advanced Communication Technology (ICACT), Pyeongchang, Korea, 19–22 February 2017; pp. 464–467.
12. Dorri, A.; Kanhere, S.S.; Jurdak, R.; Gauravaram, P. Blockchain for IoT security and privacy: The case study of a smart home. In Proceedings of the 2017 IEEE International Conference on Pervasive Computing and Communications Workshops (PerCom Workshops), Kona, HI, USA, 13–17 March 2017; pp. 618–623.
13. Zheng, Z.; Xie, S.; Dai, H.; Chen, X.; Wang, H. An overview of blockchain technology: Architecture, consensus, and future trends. In Proceedings of the 2017 IEEE International Congress on Big Data (BigData Congress), Honolulu, HI, USA, 25–30 June 2017; pp. 557–564.
14. Beikverdi, A.; Song, J. Trend of centralization in Bitcoin's distributed network. In Proceedings of the 2015 IEEE/ACIS 16th International Conference on Software Engineering, Artificial Intelligence, Networking and Parallel/Distributed Computing (SNPD), Takamatsu, Japan, 1–3 June 2015; pp. 1–6.
15. Martinez, J. Understanding Proof of Stake: The Nothing at Stake Theory. 2018. Available online: https://medium.com/coinmonks/understanding-proof-of-stake-the-nothing-at-stake-theory-1f0d71bc027 (accessed on 9 October 2019).
16. Witherspoon, Z. A Hitchhiker's Guide to Consensus Algorithms. 2017. Available online: https://hackernoon.com/a-hitchhikers-guide-to-consensus-algorithms-d81aae3eb0e3 (accessed on 9 October 2019).
17. Abadi, J.; Brunnermeier, M. *Blockchain Economics*; Technical Report; National Bureau of Economic Research: Cambridge, MA, USA, 2018.
18. Sikorski, J.J.; Haughton, J.; Kraft, M. Blockchain technology in the chemical industry: Machine-to-machine electricity market. *Appl. Energy* **2017**, *195*, 234–246. [CrossRef]
19. Weber, I.; Xu, X.; Riveret, R.; Governatori, G.; Ponomarev, A.; Mendling, J. Untrusted business process monitoring and execution using blockchain. In *International Conference on Business Process Management*; Springer: Berlin/Heidelberg, Germany, 2016; pp. 329–347.
20. Khaqqi, K.N.; Sikorski, J.J.; Hadinoto, K.; Kraft, M. Incorporating seller/buyer reputation-based system in blockchain-enabled emission trading application. *Appl. Energy* **2018**, *209*, 8–19. [CrossRef]
21. Schöner, M.M.; Kourouklis, D.; Sandner, P.; Gonzalez, E.; Förster, J. *Blockchain Technology in the Pharmaceutical Industry*; Frankfurt School Blockchain Center: Frankfurt, Germany, 2017.
22. Sylim, P.; Liu, F.; Marcelo, A.; Fontelo, P. Blockchain technology for detecting falsified and substandard drugs in distribution: Pharmaceutical supply chain intervention. *JMIR Res. Protoc.* **2018**, *7*, e10163. [CrossRef]
23. Galvez, J.F.; Mejuto, J.; Simal-Gandara, J. Future challenges on the use of blockchain for food traceability analysis. *TrAC Trends Anal. Chem.* **2018**, *107*, 222–232. [CrossRef]
24. Kamath, R. Food traceability on blockchain: Walmart's pork and mango pilots with IBM. *J. Br. Blockchain Assoc.* **2018**, *1*, 3712. [CrossRef]
25. Yue, X.; Wang, H.; Jin, D.; Li, M.; Jiang, W. Healthcare data gateways: Found healthcare intelligence on blockchain with novel privacy risk control. *J. Med. Syst.* **2016**, *40*, 218. [CrossRef]
26. Mettler, M. Blockchain technology in healthcare: The revolution starts here. In Proceedings of the 2016 IEEE 18th International Conference on e-Health Networking, Applications and Services (Healthcom), Munich, Germany, 14–17 September 2016; pp. 1–3.
27. Ekblaw, A.; Azaria, A.; Halamka, J.D.; Lippman, A. A Case Study for Blockchain in Healthcare: "MedRec" prototype for electronic health records and medical research data. In Proceedings of the IEEE Open & Big Data Conference, Vienna, Austria, 22–24 August 2016; Volume 13, p. 13.
28. Turkanović, M.; Hölbl, M.; Košič, K.; Heričko, M.; Kamišalić, A. EduCTX: A blockchain-based higher education credit platform. *IEEE Access* **2018**, *6*, 5112–5127. [CrossRef]
29. Grech, A.; Camilleri, A.F. Blockchain in Education. 2017. Available online: https://www.pedocs.de/volltexte/2018/15013/pdf/Grech_Camilleri_2017_Blockchain_in_Education.pdf (accessed on 19 April 2022).
30. Funk, E.; Riddell, J.; Ankel, F.; Cabrera, D. Blockchain technology: A data framework to improve validity, trust, and accountability of information exchange in health professions education. *Acad. Med.* **2018**, *93*, 1791–1794. [CrossRef]
31. Pop, C.; Cioara, T.; Antal, M.; Anghel, I.; Salomie, I.; Bertoncini, M. Blockchain based decentralized management of demand response programs in smart energy grids. *Sensors* **2018**, *18*, 162. [CrossRef]
32. Imbault, F.; Swiatek, M.; De Beaufort, R.; Plana, R. The green blockchain: Managing decentralized energy production and consumption. In Proceedings of the 2017 IEEE International Conference on Environment and Electrical Engineering and 2017 IEEE Industrial and Commercial Power Systems Europe (EEEIC/I&CPS Europe), Milan, Italy, 6 June 2017; pp. 1–5.

33. Aitzhan, N.Z.; Svetinovic, D. Security and privacy in decentralized energy trading through multi-signatures, blockchain and anonymous messaging streams. *IEEE Trans. Dependable Secur. Comput.* **2018**, *15*, 840–852. [CrossRef]
34. Mengelkamp, E.; Gärttner, J.; Rock, K.; Kessler, S.; Orsini, L.; Weinhardt, C. Designing microgrid energy markets: A case study: The Brooklyn Microgrid. *Appl. Energy* **2018**, *210*, 870–880. [CrossRef]
35. Pichler, M.; Meisel, M.; Goranovic, A.; Leonhartsberger, K.; Lettner, G.; Chasparis, G.; Vallant, H.; Marksteiner, S.; Bieser, H. Decentralized Energy Networks Based on Blockchain: Background, Overview and Concept Discussion. In Proceedings of the International Conference on Business Information Systems, Colorado Springs, CO, USA, 8–10 June 2018; Springer: Berlin/Heidelberg, Germany, 2018; pp. 244–257.
36. Pylon Network Team. Pylon Network Whitepaper. The Energy Blockchain Platform. 2018. Available online: https://pylon-network.org/wp-content/uploads/2019/02/WhitePaper_PYLON_v2_ENGLISH-1.pdf (accessed on 6 November 2019).
37. Suncontract. Suncontract Whitepaper. An Energy Trading Platform that Utilises Blockchain Technology to Create A New Disruptive Model for Buying and Selling Electricity. 2017. Available online: https://suncontract.org/wp-content/uploads/2020/12/whitepaper.pdf (accessed on 6 November 2019).
38. Aliyev, N.; Brooks, S.; Hale, M.; Hoy, S. Enosi Green Paper 2018. Available online: https://github.com/enosi/green-paper/blob/master/enosi-green-paper.pdf (accessed on 19 April 2022).
39. Goranović, A.; Meisel, M.; Fotiadis, L.; Wilker, S.; Treytl, A.; Sauter, T. Blockchain applications in microgrids an overview of current projects and concepts. In Proceedings of the IECON 2017-43rd Annual Conference of the IEEE Industrial Electronics Society, Beijing, China, 28 October–1 November 2017; pp. 6153–6158.
40. Koirala, B.P.; Koliou, E.; Friege, J.; Hakvoort, R.A.; Herder, P.M. Energetic communities for community energy: A review of key issues and trends shaping integrated community energy systems. *Renew. Sustain. Energy Rev.* **2016**, *56*, 722–744. [CrossRef]
41. Long, C.; Wu, J.; Zhou, Y.; Jenkins, N. Peer-to-peer energy sharing through a two-stage aggregated battery control in a community Microgrid. *Appl. Energy* **2018**, *226*, 261–276. [CrossRef]
42. Liu, N.; Yu, X.; Wang, C.; Li, C.; Ma, L.; Lei, J. Energy-sharing model with price-based demand response for microgrids of peer-to-peer prosumers. *IEEE Trans. Power Syst.* **2017**, *32*, 3569–3583. [CrossRef]
43. van Leeuwen, G.; AlSkaif, T.; Gibescu, M.; van Sark, W. An integrated blockchain-based energy management platform with bilateral trading for microgrid communities. *Appl. Energy* **2020**, *263*, 114613. [CrossRef]
44. Noor, S.; Yang, W.; Guo, M.; van Dam, K.H.; Wang, X. Energy Demand Side Management within micro-grid networks enhanced by blockchain. *Appl. Energy* **2018**, *228*, 1385–1398. [CrossRef]
45. European Parliament and Council: Regulation on the Protection of Natural Persons with Regard to the Processing of Personal Data and on the Free Movement of Such Data, and Repealing Directive 95/46/EC (Data Protection Directive). 2016. Available online: https://eur-lex.europa.eu/legal-content/EN/TXT/PDF/?uri=CELEX:32016R0679&from=EN (accessed on 9 December 2021).
46. Francisco, M.; Mezquita, Y.; Revollar, S.; Vega, P.; De Paz, J. Multi-agent distributed model predictive control with fuzzy negotiation. *Expert Syst. Appl.* **2019**, *129*, 68–83. [CrossRef]
47. Mezquita, Y.; Casado, R.; Gonzalez-Briones, A.; Prieto, J.; Corchado, J.M.; AETiC, A. Blockchain technology in IoT systems: Review of the challenges. *Ann. Emerg. Technol. Comput.* **2019**, *3*, 17–24. [CrossRef]
48. Combi, C. What Are Blockchain Confirmations and Why Do They Matter? 2017. Available online: https://coincentral.com/blockchain-confirmations/ (accessed on 19 April 2022).
49. Gatteschi, V.; Lamberti, F.; Demartini, C.; Pranteda, C.; Santamaría, V. Blockchain and smart contracts for insurance: Is the technology mature enough? *Future Internet* **2018**, *10*, 20. [CrossRef]
50. Van Saberhagen, N. CryptoNote v 2.0. 2013. Available online: https://www.getmonero.org/ru/resources/research-lab/pubs/whitepaper_annotated.pdf (accessed on 19 April 2022).
51. Roy Walker. The Battle for Blockchain Privacy: Monero. 2018. Available online: https://medium.com/all-things-venture-capital/privacy-protocol-analysis-monero-c116d7c2106f (accessed on 19 April 2022).
52. Peng, Z.; Xu, C.; Wang, H.; Huang, J.; Xu, J.; Chu, X. P2b-trace: Privacy-preserving blockchain-based contact tracing to combat pandemics. In Proceedings of the 2021 International Conference on Management of Data, Xi'an, China, 20–25 June 2021; pp. 2389–2393.
53. Bünz, B.; Bootle, J.; Boneh, D.; Poelstra, A.; Wuille, P.; Maxwell, G. Bulletproofs: Short proofs for confidential transactions and more. In Proceedings of the 2018 IEEE Symposium on Security and Privacy (SP), San Francisco, CA, USA, 21–23 May 2018; pp. 315–334.
54. Peng, Z.; Xu, J.; Chu, X.; Gao, S.; Yao, Y.; Gu, R.; Tang, Y. VFChain: Enabling verifiable and auditable federated learning via blockchain systems. *IEEE Trans. Netw. Sci. Eng.* **2021**, *9*, 173–186. [CrossRef]
55. Wu, H.; Peng, Z.; Guo, S.; Yang, Y.; Xiao, B. VQL: Efficient and Verifiable Cloud Query Services for Blockchain Systems. *IEEE Trans. Parallel Distrib. Syst.* **2021**, *33*, 1393–1406. [CrossRef]

MDPI

St. Alban-Anlage 66

4052 Basel

Switzerland

Tel. +41 61 683 77 34

Fax +41 61 302 89 18

www.mdpi.com

Energies Editorial Office

E-mail: energies@mdpi.com

www.mdpi.com/journal/energies